The Potential of Fields in Einstein's Theory of Gravitation

Zafar Ahsan

The Potential of Fields
in Einstein's Theory
of Gravitation

 Springer

Zafar Ahsan
Department of Mathematics
Aligarh Muslim University
Aligarh, Uttar Pradesh, India

ISBN 978-981-13-8978-8   ISBN 978-981-13-8976-4   (eBook)
https://doi.org/10.1007/978-981-13-8976-4

This Springer imprint is published by the registered company Springer Nature Singapore Pte Ltd.
The registered company address is: 152 Beach Road, #21-01/04 Gateway East, Singapore 189721, Singapore

Preface

The theory of relativity has developed in two phases—special theory of relativity and general theory of relativity. Special theory of relativity adapted the concept of inertial frame to the basic law of constancy of the velocity of light dispensing with the concept of absolute space and time of Galilean–Newtonian mechanics, while the general theory of relativity came into existence as an extension of the special theory of relativity.

With the tools of Riemannian geometry, Einstein was able to formulate a theory that predicts the behaviour of objects in the presence of gravitational, electromagnetic and other forces. Through his general theory of relativity, Einstein redefined gravity. From the classical point of view, gravity is the attractive force between massive objects in three-dimensional space. In general relativity, gravity manifests as a curvature of four-dimensional spacetime. Conversely, curved space and time generates effects that are equivalent to gravitational effects. J. A. Wheelar has described the results as 'matter tells spacetime how to bend and spacetime returns the complement by telling matter how to move'. On other hand, cosmology is the science of the universe as a whole, and the study of cosmology requires several kinds of physics. Since the dominant force on the cosmic scale is gravitation, this is the basic ingredient that is taken care of by Einstein's general theory of relativity. The matter distribution of fluids, gases, fields, etc., in the spacetime is given by the Einstein field equations. A cosmological model is a model of our universe that predicts the observed properties of the universe and explains the phenomena of the early universe. In a more restricted sense, cosmological models are the exact solutions of the Einstein field equations for a perfect fluid.

The global geometry of the spacetime is determined by the Riemann curvature tensor, which can be decomposed in terms of the Weyl conformal tensor, Ricci tensor and metric tensor. This decomposition involves certain irreducible tensors. In empty spacetime, the pure gravitational radiation field is described by the Weyl conformal tensor. However, when gravitational waves propagate through matter, the Weyl conformal tensor is still pertinent. In 1962, C. Lanczos thought that the Weyl conformal tensor can also be derived from a simpler tensor field. Moreover, it is known that an electromagnetic field is generated through the covariant differentiation of a vector field. The question now is: Whether it is possible or not to generate the gravitational field through a potential? The answer is yes; one can generate the gravitational field through the covariant differentiation of a tensor field. This tensor field that can act as a potential to the gravitational field is now known as the Lanczos potential. Further, certain physical problems in general relativity are often conveniently described using a tetrad formalism adapted to the geometry of the particular situation.

The present book deals with a detailed study of the Lanczos potential in general relativity and comprises eight chapters. Chapters 1–3 deal with a detailed study of tetrad formalism and its important examples—Newman–Penrose and Geroch–Held–Penrose formalisms. These discussions will then be used in the study of the Lanczos potential. Chapter 4 defines the Lanczos potential, and the equation by means of which the gravitational field is created has been derived. Such equation is called Weyl–Lanczos equation, and this equation and other related results are expressed in terms of Newman–Penrose and Geroch–Held–Penrose formalisms. Chapter 5 gives a general prescription on how to generate a gravitational field of algebraically special fields, which is supported by a number of examples, while Chap. 6 deals with yet another method to obtain the Lanczos potential for a perfect fluid spacetime, and these results are then used to generate the gravitational field of some cosmological models. Chapter 7 defines the Lanczos potentials for some well-known solutions of Einstein field equations, which have been obtained using tetrad formalisms. Apart from tetrad formalism, there are also some other methods to obtain the Lanczos potential. Such methods have been discussed in this chapter and applied to find the Lanczos potential for Gödel cosmological model. Chapter 8, contains some more applications of the Newman–Penrose formalism. A method for finding the solution of Einstein–Maxwell equations, using NP formalism, has been discussed in detail. The interaction between a Petrov type N gravitational field and null electromagnetic field has been considered, and a metric describing this situation has been obtained. A systematic and detailed study of symmetries of the spacetime (which are also known as collineations) has also been made in this chapter. Each chapter of this book ends with a list of references which by no means is a complete bibliography of the Lanczos potential and tetrad formalism; only the work referred to in this book has been included in the list.

Some portions of this book were completed at Universiti Sains Islam Malaysia (USIM), Nilai, Malaysia, during my stay as Visiting Professor. I am highly thankful to Prof. Musa Ahmad, Vice Chancellor of the university, and Dr. Nurul Sima, Head, Department of Mathematics, for their excellent support and encouragement. Thanks are also due to the learned referee for his valuable suggestions and comments. I am also grateful to my publisher for having faith in me and specially to Mr. Shamim Ahmad for introducing me to the world of Springer and his guidance during the preparation of the manuscript.

Aligarh, India   Zafar Ahsan
April 2019

Contents

About the Author

Zafar Ahsan is a former Professor at the Department of Mathematics, Aligarh Muslim University, India, where he completed his Ph.D. in Mathematics in 1979. He has previously served as a Visiting Associate at the Inter-University Centre of Astronomy and Astrophysics (IUCAA), Pune (1996–2005); a UGC Visiting Professor at Sardar Patel University, Vallabh Vidyanagar, Anand (2004–2005); and as a Visiting Professor at Universiti Sains Islam Malaysia (October–December, 2016). At present, Prof. Ahsan is a Research Fellow at Institut Sains Islam, Malaysia.

Professor Ahsan is a life member of several learned societies including the Indian Association of General Relativity and Gravitation, the Astronomical Society of India, International Electronic Engineering Mathematical Society, Indian Mathematical Society and the Tensor Society of India. Further, he has served as Editor-in-Chief of the *Aligarh Bulletin of Mathematics* (2012–2015), Editor-in-Chief of the *Journal of Tensor Society of India* (2010–2012) and Managing Editor of the *Aligarh Bulletin of Mathematics* (1998–2012). He is currently Editor, *Palestine Journal of Mathematics* and *Journal of Interpolation & Approximation in Scientific Computing*; and a member of the editorial boards of several journals, including the *Global Journal of Advanced Research on Classical and Modern Geometries*, the *Islamic University of Gaza Journal of Natural and Engineering Studies*, and the *Bulletin of the Calcutta Mathematical Society* and the *Journal of the Calcutta Mathematical Society*. With over 40 years of teaching experience, he has published five books and over 100 research papers in several international journals of repute. His research interests are in gravitational waves, symmetries of space–time, exact solutions of Einstein equations, tetrad formalisms and differential geometric structures in general relativity. His current research interest, apart from General Relativity and Gravitation, is in the Qur'an, Science and Modern Cosmology.

Professor Ahsan is the President of Anjuman Faroogh-e-Science (Association for Promotion of Science), Aligarh branch; and has been a Council Member of the Indian Association for General Relativity and Gravitation (2002–2006). He is the recipient of the Nishan-e-Azad award (2017) for the promotion of science in Urdu; the International Einstein Award for Scientific Achievement (2011); the Rashtriya

Gaurav Award (2007); the Shiksha Rattan Puraskar (2007); and the Best Citizens of India award (2006, 2007, 2008). He is a Member of the Research Board of Advisers, American Biography Institute (USA) and the Life Time President of the Tensor Society of India.

Chapter 1
The Tetrad Formalism

1.1 Introduction

The standard way of treating problems in general theory of relativity is to consider Einstein field equations in a local coordinate basis adapted to the problems with which one is working. In recent times, it has proved advantageous to choose a suitable basis of four linearly independent vectors, to project the relevant quantities on to the chosen basis and consider the equations satisfied by them. This is *tetrad formalism*—and in general relativity certain types of calculations are conveniently carried out if one deals with a tetrad formalism [2]. This chapter deals with the study of such formalism which will then be used to develop the Newman–Penrose and Geroch–Held–Penrose formalisms in subsequent chapters.

1.2 The Tetrad Representation

At each point of the spacetime, set a basis of four contravariant vectors

$$e_{(a)}{}^i \quad (a = 1, 2, 3, 4) \tag{1.1}$$

where the indices enclosed in parenthesis, such as (a), (b), (c), (d), etc., are the terad indices and i, j, k, l, etc., are the tensor indices. From Eq. (1.1) and the metric tensor g_{ij}, the covariant vector is

$$e_{(a)i} = g_{ik} e_{(a)}{}^k \tag{1.2}$$

Also

$$e_{(a)}^i e_i^{(b)} = \delta_{(a)}^{(b)}, \ e_{(a)}^i e_j^{(a)} = \delta_j^i \tag{1.3}$$

© Springer Nature Singapore Pte Ltd. 2019
Z. Ahsan, *The Potential of Fields in Einstein's Theory of Gravitation*,
https://doi.org/10.1007/978-981-13-8976-4_1

where $e_i^{(b)}$ is the inverse of the matrix $[e_{(a)}^i]$ (with the tetrad index labelling the rows and the tensor index labelling the columns). Further assume that

$$e_{(a)}^i e_{(b)i} = \eta_{(a)(b)} \tag{1.4}$$

where

$$\eta_{(a)(b)} \text{ is a constant symmetric matrix} \tag{1.5}$$

If $\eta^{(a)(b)}$ is the inverse of the matrix $[\eta_{(a)(b)}]$, then

$$\eta^{(a)(b)} \eta_{(b)(c)} = \delta_{(c)}^{(a)} \tag{1.6}$$

Also

$$\eta_{(a)(b)} e_i^{(a)} = e_{(b)i} \tag{1.7a}$$

$$\eta^{(a)(b)} e_{(a)i} = e_i^{(b)} \tag{1.7b}$$

$$e_{(a)i} e_j^{(a)} = g_{ij} \tag{1.8}$$

To obtain the tetrad components of a given vector or tensor field, we project it onto the tetrad frame and thus [1, 2]

$$A_{(a)} = e_{(a)j} A^j = e_{(a)}^j A_j$$

$$A^{(a)} = \eta^{(a)(b)} A_{(b)} = e_j^{(a)} A^j = e^{(a)j} A_j \tag{1.9a}$$

$$A^i = e_{(a)}^i A^{(a)} = e^{(a)i} A_{(a)}$$

$$A_i = e_i^{(a)} A_{(a)} = e_{(a)i} A^{(a)}$$

and

$$T_{(a)(b)} = e_{(a)}^i e_{(b)}^j T_{ij} = e_{(a)}^i T_{i(b)}$$

$$T^{(a)(b)} = e_{,i}^{(a)} e_j^{(b)} T^{ij} = e_i^{(a)} T^{i(b)} \tag{1.9b}$$

$$T_{ij} = e_i^{(a)} e_j^{(b)} T_{(a)(b)} = e_i^{(a)} T_{(a)j}$$

$$T^{ij} = e_{(a)}^i e_{(b)}^j T^{(a)(b)} = e_{(a)}^i T^{(a)j}$$

Remarks

1. If in the above considerations, the basis vectors are orthonormal, then

$$\eta_{(a)(b)} = \text{diag} \, (\, 1, \, -1, \, -1, \, -1 \,) = \begin{bmatrix} 1 & 0 & 0 & 0 \\ 0 & -1 & 0 & 0 \\ 0 & 0 & -1 & 0 \\ 0 & 0 & 0 & -1 \end{bmatrix}$$

2. (i) $\eta^{(a)(b)}$ and $\eta_{(a)(b)}$ can be used to raise and lower the tetrad indices even as the tensor indices are raised and lowered with the metric tensor.
(ii) There is no ambiguity in having quantities in which both tetrad and tensor indices occur.
(iii) The result of contracting a tensor is the same whether it is carried out with respect to its tensor or tetrad indices.

1.3 Directional Derivative and Ricci Rotation Coefficient

In general, a vector cannot be considered as an arrow connecting two points of the manifold. To have consistent generalization of the concept of vectors in E^n, we identify vectors on the manifold M with tangent vectors.

Definition 1.1 A *tangent vector* **v** at p is an operator which assigns to each differentiable function on M a real number **v(f)**. This operator satisfies
(a) $\mathbf{v(f + h)} = \mathbf{v(f)} + \mathbf{v(h)}$.
(b) $\mathbf{v(fh)} = \mathbf{hv(f)} + \mathbf{fv(h)}$.
(c) $\mathbf{v(cf)} = \mathbf{cv(f)}$, where c is a constant.
Thus, $\mathbf{v(c)} = \mathbf{0}$ for any constant function c. The above definition is independent of the choice of the coordinates. A tangent vector is just the directional derivative along a curve $\gamma(t)$ through p [A smooth curve $\gamma(t)$ on the manifold M is defined by a differentiable map of an interval of the real line into $M, \gamma(t) : -\epsilon < t < \epsilon \rightarrow M$].

Using the Taylor series expansion for the function f at p and using the Definition 1.1, the tangent vector **v** at p can be written as

$$\mathbf{v} = v_i \, \frac{\partial}{\partial x^i} \tag{1.10}$$

The real coefficients v_i are the components of **v** at p with respect to the local coordinate system $(x^1, x^2, \ldots, x^n)$ in a neighbourhood of p. According to Eq. (1.10), the directional derivative along the coordinate lines at p forms a basis of an n-dimensional vector space whose elements are the tangent vectors at p. This space is called the *tangent space T_p*. The basis $\{\frac{\partial}{\partial x^i}\}$ is called a *coordinate basis* or *holonomic frame*. We also have

Definition 1.2 The contravariant vector $\mathbf{e}_{(a)}$, considered as the tangent vector, defines the *directional derivative*

$$\mathbf{e}_{(a)} = e^i_{(a)} \frac{\partial}{\partial x^i} \tag{1.11}$$

and we write

$$\phi_{,a} = e^i_{(a)} \frac{\partial \phi}{\partial x^i} = e^i_{(a)} \phi_{,i} \tag{1.12}$$

[the tangent vector/directional derivative defined by Eq. (1.10) leads to Eq. (1.11)], where ϕ is any scalar field, a comma indicates the partial differentiation and the semicolon denotes the covariant differentiation.

In general, we define

$$A_{(a),(b)} = e^i_{(b)} \frac{\partial}{\partial x^i} A_{(a)} = e^i_{(b)} \frac{\partial}{\partial x^i} \left[e^j_{(a)} A_j \right] = e^i_{(b)} \nabla_{\partial_i} \left[e^j_{(a)} A_j \right]$$

$$= e^i_{(b)} \left[e^k_{(a);j} A_k + e^j_{(a)} A_{j;i} \right] \tag{1.13}$$

$$= e^j_{(a)} A_{j;i} e^i_{(b)} + e^i_{(b)} e^k_{(a);j} A_k = e^j_{(a)} A_{j;i} e^i_{(b)} + e_{(a)k;i} e^i_{(b)} e^k_{(c)} A^{(c)} \tag{1.14}$$

(According to $\nabla_{\bar{X}} f = \bar{X} = X^i \partial f / \partial x^i$ it may be noted that in a local coordinate basis, ∂_k and ∇_{∂_k}, when acting on functions coincides with partial differentiation with respect to k.)

We define

$$\gamma_{(c)(a)(b)} = e^k_{(c)} e_{(a)k;i} e^i_{(b)} \tag{1.15}$$

as *Ricci rotation coefficients* and Eq. (1.14) can now be written as

$$A_{(a),(b)} = e^j_{(a)} A_{j;i} e^i_{(b)} + \gamma_{(c)(a)(b)} A^{(c)} \tag{1.16}$$

The Ricci rotation coefficients given by Eq. (1.15) can equivalently be defined as

$$e_{(a)k;i} = e^{(c)}_k \gamma_{(c)(a)(b)} e^{(b)}_i \tag{1.17}$$

Note. Ricci rotation coefficients are skew-symmetric in the first pair of indices:

$$\gamma_{(c)(a)(b)} = -\gamma_{(a)(c)(b)} \tag{1.18}$$

[This can be proved by the expansion of the identity

$$0 = \eta_{(a)(b),i} = [e_{(a)k} e^k_{(b)}]_{;i}] \tag{1.19}$$

From Eq. (1.17), we have

$$e_{(a)k;i} = e_k^{(c)} \, \gamma_{(c)(a)(b)} \, e_i^{(b)} = \gamma_{k(a)i} \tag{1.20}$$

which due to skew-symmetry of the Ricci rotation coefficients leads to

$$e_{(a);i}^k = -\gamma_{(a)i}^k \tag{1.21}$$

From Eq. (1.16), we have

$$e_{(a)}^j \, A_{j;i} \, e_{(b)}^i = A_{(a),(b)} - \gamma_{(c)(a)(b)} A^{(c)} = A_{(a),(b)} - \eta^{(n)(m)} \gamma_{(n)(a)(b)} A^{(m)} = A_{(a)|(b)} \tag{1.22}$$

where

$$A_{(a)|(b)} = A_{(a),(b)} - \eta^{(n)(m)} \, \gamma_{(n)(a)(b)} \, A^{(m)} \tag{1.23}$$

is called the *intrinsic derivative* of $A_{(a)}$ in the direction of $e_{(b)}$.

From Eq. (1.22), we have

$$A_{i;j} = e_i^{(a)} \, A_{(a)|(b)} \, e_j^{(b)} \tag{1.24}$$

which gives the relationship between directional (covariant) and intrinsic derivatives.

From Eq. (1.22), the intrinsic derivative of the Riemann curvature tensor is

$$R_{(a)(b)(c)(d) \, | \, (f)} = R_{ijkl;m} \, e_{(a)}^i \, e_{(b)}^j \, e_{(c)}^k \, e_{(d)}^l \, e_{(f)}^m \tag{1.25}$$

which can alternatively be written as

$$R_{(a)(b)(c)(d) \, | \, (f)} = [R_{ijkl} \, e_{(a)}^i \, e_{(b)}^j \, e_{(c)}^k \, e_{(d)}^l]_{;m} \, e_{(f)}^m$$

$$= R_{ijkl;m} \, e_{(a)}^i \, e_{(b)}^j \, e_{(c)}^k \, e_{(d)}^l \, e_{(f)}^m + R_{ijkl} \, e_{(a);m}^i \, e_{(b)}^j \, e_{(c)}^k \, e_{(d)}^l \, e_{(f)}^m$$

$$+ R_{ijkl} \, e_{(a)}^i \, e_{(b);m}^j \, e_{(c)}^k \, e_{(d)}^l \, e_{(f)}^m + R_{ijkl} \, e_{(a)}^i \, e_{(b)}^j \, e_{(c);m}^k \, e_{(d)}^l \, e_{(f)}^m$$

$$+ R_{ijkl} \, e_{(a)}^i \, e_{(b)}^j \, e_{(c)}^k \, e_{(d);m}^l \, e_{(f)}^m$$

Now using Eqs. (1.21), (1.9), (1.10) and (1.7) (for raising/lowering the tetrad indices), the above equation reduces to

$$R_{(a)(b)(c)(d) \, | \, (f)} = R_{(a)(b)(c)(d),(f)}$$

$$-\eta^{(p)(q)} \, [\gamma_{(p)(a)(f)} \, R_{(q)(b)(c)(d)} + \gamma_{(p)(b)(f)} \, R_{(a)(q)(c)(d)}$$

$$+ \gamma_{(p)(c)(f)} \, R_{(a)(b)(q)(d)} + \gamma_{(p)(d)(f)} \, R_{(a)(b)(c)(q)}] \tag{1.26}$$

which is analogous to Eq. (1.23).

We shall now prove the following lemma.

Lemma 1.1 *The evaluation of Ricci rotation coefficients does not involve the eval-uation of covariant derivatives (and thus the evaluation of Christoffel symbols is not required).*

Proof Define a symbol $\lambda_{(a)(b)(c)}$ which is skew-symmetric in the first and last index as

$$\lambda_{(a)(b)(c)} = e_{(b)i,j} [e^i_{(a)} e^j_{(c)} - e^j_{(a)} e^i_{(c)}] \tag{1.27}$$

which can be written as

$$\lambda_{(a)(b)(c)} = [e_{(b)i,j} - e_{(b)j,i}] e^i_{(a)} e^j_{(c)}$$

But from equation

$$f_{,j;k} - f_{,k;j} = - f_{,l} (\Gamma^l_{jk} - \Gamma^l_{kj})$$

we can replace the ordinary (partial) derivatives of $e_{(b)i}$ and $e_{(b)j}$ by the covariant derivatives and we have

$$\lambda_{(a)(b)(c)} = \left[e_{(b)i;j} - e_{(b)j;i} \right] e^i_{(a)} e^j_{(c)} = e_{(b)i;j} e^i_{(a)} e^j_{(c)} - e_{(b)j;i} e^i_{(a)} e^j_{(c)}$$

$$= e^i_{(a)} e_{(b)i;j} e^j_{(c)} - e^j_{(c)} e_{(b)j;i} e^i_{(a)} = \gamma_{(a)(b)(c)} - \gamma_{(c)(b)(a)} \tag{1.28}$$

From Eq. (1.28), we have

$$\gamma_{(a)(b)(c)} = \frac{1}{2} \left[\lambda_{(a)(b)(c)} + \lambda_{(c)(a)(b)} - \lambda_{(b)(c)(a)} \right] \tag{1.29}$$

Also from Eq. (1.27) we see that $\lambda_{(a)(b)(c)}$ does not involve Christoffel symbols and thus Eq. (1.29) does not require the calculation of Christoffel symbols and hence the evaluation of Ricci rotation coefficients does not involve the evaluation of covariant derivatives.

It may be noted that

$$\lambda_{(a)(b)(c)} = - \lambda_{(c)(b)(a)} \tag{1.30}$$

1.4   The Commutation Relation and the Structure Constants

Given any two vector fields X, Y, their Lie bracket is defined as

$$[X, Y] f = (XY - YX) f = X(Yf) - Y(Xf) \tag{1.31a}$$

and

$$\mathcal{L}_X Y = [X, Y] = -[Y, X] = -\mathcal{L}_Y X \tag{1.31b}$$

$$[X, Y]^j = X^k Y^j_{,k} - Y^k X^j_{,k} \text{ (in local coordinates)} \tag{1.31c}$$

where $\mathcal{L}_X Y$ denotes the Lie derivative of Y with respect to X. Consider the basis $e_{(a)}$, we have

$$[e_{(a)}, e_{(b)}] = C_{(a)(b)}{}^{(c)} e_{(c)}, \quad C_{(a)(b)}{}^{(c)} = -C_{(b)(a)}{}^{(c)} \tag{1.32}$$

where $C_{(a)(b)}{}^{(c)}$ are called the *structure constants* and are 24 in number. Also

$$[e_{(a)}, e_{(b)}] f = e^i_{(a)} [e^j_{(b)} f_{,j}]_{,i} - e^i_{(b)} [e^j_{(a)} f_{,j}]_{,i}$$

$$= [e^i_{(a)} e^j_{(b),i} - e^i_{(b)} e^j_{(a),i}] f_{,j} = [e^i_{(a)} e^j_{(b);i} - e^i_{(b)} e^j_{(a);i}] f_{,j}$$

$$= [-\gamma^{(c)}_{(b)(a)} + \gamma^{(c)}_{(a)(b)}] e^j_{(c)} f_{,j} \tag{1.33}$$

(from the definition of Ricci rotation coefficients). From Eqs. (1.32) and (1.33), we have

$$C_{(a)(b)}{}^{(c)} = \gamma^{(c)}_{(b)(a)} - \gamma^{(c)}_{(a)(b)} \tag{1.34}$$

Equation (1.34) provides *commutation relations*. There are 24 commutation relations.

1.5 The Ricci and the Bianchi Identities

We know that the Riemann tensor is defined through the Ricci identity

$$R^i_{jkl} A_i = A_{j;k;l} - A_{j;l;k}$$

or

$$R_{hijk} A^h = A_{i;j;k} - A_{i;k;j}$$

and thus

$$R_{hijk} e^{(h)}_{(a)} = e_{(a)i;j;k} - e_{(a)i;k;j} \tag{1.35}$$

The tetrad representation of R_{hijk}, using Eq. (1.9b), is given by

$$R_{(a)(b)(c)(d)} = R_{hijk} e^h_{(a)} e^i_{(b)} e^j_{(c)} e^k_{(d)} = [e_{(a)i;j;k} - e_{(a)i;k;j}] e^i_{(b)} e^j_{(c)} e^k_{(d)}$$

which from Eq. (1.17) reduces to

$$R_{(a)(b)(c)(d)} = \left\{ \left[e_i^{(f)} \, \gamma_{(f)(a)(g)} \, e_j^{(g)} \right]_{;k} - \left[e_i^{(f)} \, \gamma_{(f)(a)(g)} \, e_k^{(g)} \right]_{;j} \right\} e_{(b)}^i \, e_{(c)}^j \, e_{(d)}^k$$

Using Eq. (1.18), after simplification, we get

$$R_{(a)(b)(c)(d)} = - \gamma_{(a)(f)(g),k} \, e_i^{(f)} \, e_j^{(g)} \, e_{(b)}^i \, e_{(c)}^j \, e_{(d)}^k + \gamma_{(a)(f)(g),j} \, e_i^{(f)} \, e_k^{(g)} \, e_{(b)}^i \, e_{(c)}^j \, e_{(d)}^k$$

$$+ \gamma_{(f)(a)(g)} \{ e_{i;k}^{(f)} \, e_j^{(g)} + e_i^{(f)} \, e_{j;k}^{(g)} \} e_{(b)}^i \, e_{(c)}^j \, e_{(d)}^k$$

$$+ \gamma_{(f)(a)(g)} \{ e_{i;j}^{(f)} \, e_k^{(g)} + e_i^{(f)} \, e_{k;j}^{(g)} \} e_{(b)}^i \, e_{(c)}^j \, e_{(d)}^k$$

$$= - \gamma_{(a)(f)(g),k} \, e_{(d)}^k \, \eta_{(b)}^{(f)} \, \eta_{(c)}^{(g)} + \gamma_{(a)(f)(g),j} \, e_{(c)}^j \, \eta_{(b)}^{(f)} \, \eta_{(d)}^{(g)}$$

$$+ \gamma_{(f)(a)(g)} \{ e_{i;k}^{(f)} \, e_j^{(g)} - e_{i;j}^{(f)} \, e_k^{(g)} \} e_{(b)}^i \, e_{(c)}^j \, e_{(d)}^k$$

$$+ \gamma_{(f)(a)(g)} \{ e_{j;k}^{(g)} \, e_i^{(f)} - e_{k;j}^{(g)} \, e_i^{(f)} \} e_{(b)}^i \, e_{(c)}^j \, e_{(d)}^k$$

$$= - \gamma_{(a)(f)(g),k} \, e_{(d)}^k \, \eta_{(b)}^{(f)} \, \eta_{(c)}^{(g)} + \gamma_{(a)(f)(g),j} \, e_{(c)}^j \, \eta_{(b)}^{(f)} \, \eta_{(d)}^{(g)}$$

$$+ \gamma_{(f)(a)(g)} \{ \gamma_{ik}^{(f)} \, e_j^{(g)} - \gamma_{ij}^{(f)} \, e_k^{(g)} \} e_{(b)}^i \, e_{(c)}^j \, e_{(d)}^k$$

$$+ \gamma_{(f)(a)(g)} \{ \gamma_{jk}^{(g)} - \gamma_{kj}^{(g)} \} e_i^{(f)} \, e_{(b)}^i \, e_{(c)}^j \, e_{(d)}^k$$

$$= - \gamma_{(a)(b)(c),(d)} + \gamma_{(a)(b)(d),(c)} + \gamma_{(f)(a)(g)} \{ \gamma_{(b)(d)}^{(f)} \, \eta_{(c)}^{(g)} - \gamma_{(b)(c)}^{(f)} \, \eta_{(d)}^{(g)} \}$$

$$+ \gamma_{(f)(a)(g)} \{ \gamma_{(c)(d)}^{(g)} - \gamma_{(d)(c)}^{(g)} \} \eta_{(b)}^{(f)}$$

$$= - \gamma_{(a)(b)(c),(d)} + \gamma_{(a)(b)(d),(c)} + \gamma_{(b)(a)(g)} \{ \gamma_{(c)(d)}^{(g)} - \gamma_{(d)(c)}^{(g)} \}$$

$$+ \gamma_{(f)(a)(c)} \, \gamma_{(b)(d)}^{(f)} - \gamma_{(f)(a)(d)} \, \gamma_{(b)(c)}^{(f)} \tag{1.36}$$

Equation (1.36) is skew-symmetric in the pair (c) and (d). Also, since the Ricci rotation coefficients are skew-symmetric in first pair of indices, the total number of equations in Eq. (1.36) is $6 \times 6 = 36$.

The Bianchi identities

$$R_{ij[kl;m]} = \frac{1}{3} \left(R_{ijkl;m} + R_{ijlm;k} + R_{ijmk;l} \right) = 0$$

when written in terms the intrinsic derivatives and the tetrad components can be expressed as

$$R_{(a)(b)[(c)(d)|(f)]} = \frac{1}{6} \sum_{[(c)(d)(f)]} \left\{ R_{(a)(b)(c)(d),(f)} - \eta^{(n)(m)} \left[\gamma_{(n)(a)(f)} \, R_{(m)(b)(c)(d)} \right. \right.$$

$$\left. \left. + \gamma_{(n)(b)(f)} \, R_{(a)(m)(c)(d)} + \gamma_{(n)(c)(f)} \, R_{(a)(b)(m)(d)} + \gamma_{(n)(d)(f)} \, R_{(a)(b)(c)(m)} \right] \right\} \quad (1.37)$$

The Bianchi identities include 24 distinct equations corresponding to six distinct index pairs $(i, j; i \neq j)$, each of which can be associated with the four choices $k \neq l \neq m$. However, only 20 of these four equations are linearly independent.

Thus, the basic equations of the tetrad formalism are 24 commutation relations [Eq. (1.32)], *the 36 Ricci identities* [Eq. (1.36)] *and 20 linearly independent Bianchi identities given by* Eq. (1.37).

References

1. Ahsan, Z.: Lecture Notes on Tetrad Formalism: Lectures Delivered at Workshop on Geometry, Gravity and Cosmology. Sardar Patel University, Vallabh Vidyanagar (2006)
2. Chandrashekhar, S.: Mathematical Theory of Black Holes. Oxford University Press, New York (1983)

Chapter 2
The Newman–Penrose Formalism

2.1 Introduction

The Newman–Penrose formalism (also known as spin-coefficient formalism) is a tetrad formalism with special choice of the basis vector. The beauty of this formalism, when it was first proposed by Newman and Penrose in 1962 [13], was precisely in their choice of a null basis. The underlying motivation for a choice of null basis was Penrose's strong belief that the essential element of a spacetime is its light cone structure which makes possible the introduction of a spinor basis. The expanded system of equations connecting the spinor components of the Riemann curvature tensor with the components of the spinor connections (spin-coefficients) has become known as the system of Newman–Penrose equations (or briefly, NP equations). It is possible that the formalism may look somewhat cumbersome with long formulas and tedious calculations, and usually creates some psychological barrier in handling and using NP method; but once the initial hurdle is crossed, the formalism offers a deep insight into the symmetries of the spacetime.

However, from the time elapsed since its first appearance in 1962 [13], the viability and convenience of NP formalism have become established. Moreover in the modern literature, it is generally accepted and widely used. What is the reason for such popularity of NP formalism? Apparently, the main reason consists in its adequacy and internal adaptability for constructing the exact solutions of Einstein field equations and for other investigations (cf., [1–4, 10, 22]).

It is known that the Weyl tensor determines a set of principal light-like vector [or principal null direction (pnd)] the mutual position of which is directly related to the algebraic types of the gravitational fields. The integral curves of pnd form a family (congruence) of light curves on the spacetime manifold. The geometric meaning of such congruences can lead to a classification of gravitational fields. If a field of complex light tetrad is chosen so that one of the two real light-like vectors of the tetrad coincide with a principal light vector, then the individual spin-coefficient coincides with the so-called optical scalars. The Newman–Penrose (NP) equations, in such case, which relate different optical scalars among themselves, provide a completely

© Springer Nature Singapore Pte Ltd. 2019
Z. Ahsan, *The Potential of Fields in Einstein's Theory of Gravitation*,
https://doi.org/10.1007/978-981-13-8976-4_2

definite geometric meaning and are helpful in obtaining the solutions of Einstein field equations (cf., [20]).

When a complex null tetrad is used to describe the geometry of the spacetime then NP equations provide a considerable simplification in the equations. The use of a complex null tetrad and the choice of the light coordinates make the NP formalism a convenient tool for the description of massless fields (photons, neutrinos) and the gravitational fields.

The NP equations have proved fruitful in studying the asymptotic behaviour of the gravitational fields [15]. From the definition of asymptotic flatness and the use of NP system, it is possible to study the behaviour of the fields at infinity. These methods along with the idea of conformal infinity, enable Hawking to study the gravitational field of a black hole. Lind et al. [11] establish the relationship between the asymptotic properties of the gravitational field and the nature of the motion of a body generating the field. In this relationship, the integration of NP equations plays an important role. The NP formalism has other number of applications too (cf., [6–8] and Chap. 8).

Motivated by such vast applications of NP formalism, this chapter is devoted to the study of this formalism.

2.2 Complex Null Tetrad and the Spin-Coefficients

Instead of the orthonormal basis $(e_{(x)}, e_{(y)}, e_{(z)}, e_{(t)})$, Newman and Penrose have chosen a null basis comprising of a pair of real null vectors l^i and n^i and a pair of complex conjugate null vectors m^i and $\bar{m}^i$ such that

$$l^i m_i = l^i \bar{m}_i = n^i m_i = n^i \bar{m}_i = 0 \tag{2.1}$$

$$l^i l_i = n^i n_i = m^i m_i = \bar{m}^i \bar{m}_i = 0 \tag{2.2}$$

and the normalization condition

$$l^i n_i = 1, \quad m^i \bar{m}_i = -1 \tag{2.3}$$

The fundamental matrix represented by $\eta_{(a)(b)}$ is a constant symmetric matrix of the form

$$[\,\eta_{(a)(b)}\,] = [\,\eta^{(a)(b)}\,] = \begin{bmatrix} 0 & 1 & 0 & 0 \\ 1 & 0 & 0 & 0 \\ 0 & 0 & 0 & -1 \\ 0 & 0 & -1 & 0 \end{bmatrix} \tag{2.4}$$

The null basis $\{\, l^i, n^i, m^i, \bar{m}^i \,\}$ and the orthonormal basis $e^i_{(a)}$ are related through the equations

$$e_{(1)} = l^i = \frac{1}{\sqrt{2}}(e_{(t)} + e_{(z)}), \quad e_{(2)} = n^i = \frac{1}{\sqrt{2}}(e_{(t)} - e_{(z)})$$

$$e_{(3)} = m^i = \frac{1}{\sqrt{2}}(e_{(x)} + ie_{(y)}), \quad e_{(4)} = \bar{m}^i = \frac{1}{\sqrt{2}}(e_{(x)} - ie_{(y)}) \qquad (2.5)$$

The corresponding covariant basis is given by

$$e^{(1)} = e_{(2)} = n^i, \; e^{(2)} = e_{(1)} = l^i, \; e^{(3)} = -e_{(4)} = -\bar{m}^i, \; e^{(4)} = -e_{(3)} = -m^i \tag{2.6}$$

The basis vectors, regarded as directional derivatives [cf., Eq. (1.11)], can be denoted as

$$e_1 = e^2 = D = l^i \frac{\partial}{\partial x^i}, \; e_2 = e^1 = \Delta = n^i \frac{\partial}{\partial x^i}$$

$$e_3 = -e^4 = \delta = m^i \frac{\partial}{\partial x^i}, \; e_4 = -e^3 = \bar{\delta} = \bar{m}^i \frac{\partial}{\partial x^i} \qquad (2.7)$$

Once a field of complex null tetrad is assigned in spacetime, the tetrad formalism can then be used for describing geometric objects. The contravariant components of the metric tensor in terms of null tetrad are

$$g^{ij} = e^i_{(a)} e^j_{(b)} \eta^{(a)(b)} = l^i n^j + n^i l^j - m^i \bar{m}^j - \bar{m}^i m^j \tag{2.8}$$

while

$$g_{ij} = e_{(a)i} e_{(b)j} \eta^{(a)(b)} = l_i n_j + n_i l_j - m_i \bar{m}_j - \bar{m}_i m_j \tag{2.9}$$

where

$$\eta_{(a)(b)} = e_{(a)i} e_{(b)j} g^{ij} \tag{2.10}$$

The Ricci rotation coefficients [cf., Eq. (1.15)], using Eq. (2.6) leads to the following spin-coefficients:

$$\kappa = \gamma_{311} = l_{i;j} m^i l^j, \; \tau = \gamma_{312} = l_{i;j} m^i n^j$$

$$\sigma = \gamma_{313} = l_{i;j} m^i m^j, \; \rho = \gamma_{314} = l_{i;j} m^i \bar{m}^j$$

$$\pi = \gamma_{241} = -n_{i;j} \bar{m}^i l^j, \; \nu = \gamma_{242} = -n_{i;j} \bar{m}^i n^j$$

$$\mu = \gamma_{243} = -n_{i;j} \bar{m}^i m^j, \; \lambda = \gamma_{244} = -n_{i;j} \bar{m}^i \bar{m}^j \tag{2.11a}$$

$$\epsilon = \frac{1}{2}(\gamma_{211} + \gamma_{341}) = \frac{1}{2}(l_{i;j} n^i l^j - m_{i;j} \bar{m}^i l^j)$$

$$\gamma = \frac{1}{2}(\gamma_{212} + \gamma_{342}) = \frac{1}{2}(l_{i;j} n^i n^j - m_{i;j} \bar{m}^i n^j) \tag{2.11b}$$

$$\beta = \frac{1}{2}(\gamma_{213} + \gamma_{343}) = \frac{1}{2}(l_{i;j}n^i m^j - m_{i;j}\bar{m}^i m^j)$$

$$\alpha = \frac{1}{2}(\gamma_{214} + \gamma_{344}) = \frac{1}{2}(l_{i;j}n^i \bar{m}^j - m_{i;j}\bar{m}^i \bar{m}^j) \tag{2.11b}$$

It may be noted that the complex conjugate of any quantity can be obtained by replacing index 3, wherever it occurs, by index 4, and conversely (hence onward we are not using the square brackets for the tetrad indices).

2.3 The Riemann, Ricci and Weyl Tensors

The general theory of relativity is a theory of gravitation in which gravitation emerges as the property of the spacetime structure through the metric tensor g_{ij}. The metric tensor determines another object known as Riemann curvature tensor. At any given event this tensorial object provides all information about the gravitational field in the neighbourhood of the event. It may be interpreted as describing the curvature of the spacetime. The Riemann curvature tensor is the simplest non-trivial object one can build at a point; its vanishing is the criterion for the absence of genuine gravitational fields and its structure determines the relative motion of the neighbouring test particles via the equation of geodesic deviation [5].

The Riemann curvature tensor R^k_{ijl}, for a covariant vector field u_k, is defined through the Ricci identity (cf., [5])

$$R^i{}_{jkl}\, u_i = u_{j;k;l} - u_{j;l;k} \tag{2.12a}$$

where

$$R^k_{ijl} = \frac{\partial}{\partial x^j}\Gamma^k_{il} - \frac{\partial}{\partial x^l}\Gamma^k_{ij} + \Gamma^m_{il}\Gamma^k_{mj} - \Gamma^m_{ij}\Gamma^k_{ml} \tag{2.12b}$$

For this tensor, we have

$$R_{ijkl} = -R_{jikl}, \ R_{ijkl} = -R_{ijlk}, \ R_{(ij)(kl)} = R_{(kl)(ij)} \tag{2.13a}$$

$$R_{ijkl} + R_{iklj} + R_{iljk} = 0 \tag{2.13b}$$

$$R_{ijkl;m} + R_{ijlm;k} + R_{ijmk;l} = 0 \quad \text{(Bianchi identity)} \tag{2.13c}$$

The symmetries of Riemann tensor lead to a rank two tensor (by contraction) which is known as the Ricci tensor and is defined as

$$R_{ik} = g^{jl}R_{ijkl} \tag{2.14}$$

The Riemann tensor can be separated into a 'trace-free' and a 'Ricci part' and this separation is established by the Weyl tensor (in n dimension) as

$$C_{ijkl} = R_{ijkl} - \frac{1}{n-2}(g_{ik}R_{jl} + g_{jl}R_{ik} - g_{jk}R_{il} - g_{il}R_{jk})$$

$$+ \frac{1}{(n-1)(n-2)}(g_{ik}g_{jl} - g_{il}g_{jk})R \tag{2.15}$$

so that in four dimensions, the Weyl tensor is related to the Riemann and Ricci tensors through the equation

$$C_{ijkl} = R_{ijkl} - \frac{1}{2}(g_{ik}R_{jl} + g_{jl}R_{ik} - g_{jk}R_{il} - g_{il}R_{jk})$$

$$+ \frac{1}{6}(g_{ik}g_{jl} - g_{il}g_{jk})R \tag{2.16}$$

where $R = g^{ij}R_{ij}$ is the scalar curvature (cf., [5]). This tensor has all the symmetries of the Riemann tensor and also has the property that $g^{jl}C_{ijkl} = 0$ in contrast to $g^{jl}R_{ijkl} = R_{ik}$. The Riemann tensor has 20 independent components, while the Weyl and Ricci tensors have 10 independent components each.

In empty spacetime, $R_{ij} = 0$ and the Riemann tensor reduces to Weyl tensor. Thus, in order to have a classification of vacuum Riemann tensor, it is sufficient to classify the Weyl tensor. There are three main approaches for the classification of Weyl tensor (gravitational field)-namely: the matrix method [16, 17, 21], the spinor method [14, 18] and the tensor method [19]. The tensor method is equivalent to the other two [12]. The classification of the Weyl tensor in empty spacetime leads to different types of gravitational fields which are denoted as Petrov types I, II, D, III and N. Type I gravitational fields are known as algebraically general, while types II, D, III and N are algebraically special. Moreover, Petrov types III and N gravitational fields correspond to the state of gravitational radiation.

The Weyl tensor is the trace-free (i.e. the contraction with each pair of indices is zero) part of the Riemann tensor, and the relationship between the tetrad components of the Riemann, the Weyl and the Ricci tensors is given by

$$R_{abcd} = C_{abcd} - \frac{1}{2}(\eta_{ac}R_{bd} + \eta_{bd}R_{ac} - \eta_{bc}R_{ad} - g_{ad}R_{bc})$$

$$+ \frac{1}{6}(\eta_{ac}\eta_{bd} - \eta_{ad}\eta_{bc})R \tag{2.17}$$

where R_{bd} denotes the tetrad components of the Ricci tensor and R, the Ricci scalar curvature:

$$R_{ac} = \eta^{bd}R_{abcd}, \quad R = \eta^{ab}R_{ab} = 2(R_{12} - R_{34}) \tag{2.18}$$

Since C_{abcd} is trace-free (there are only 10 real components of the Weyl tensor)

$$\eta^{ad} C_{abcd} = 0 = C_{1bc2} + C_{2bc1} - C_{3bc4} - C_{4bc3} \qquad (2.19)$$

and the cyclic condition leads to

$$C_{1234} + C_{1342} + C_{1423} = 0 \qquad (2.20)$$

Take $b = c$ ($b, c = 1, 2, 3, 4$) in Eq. (2.19), we get

$$C_{1314} = C_{2324} = C_{1332} = C_{1442} = 0 \qquad (2.21)$$

while taking $b \neq c$ in Eq. (2.19), using Eq. (2.20), we get

$$C_{1231} = C_{1334}, \ \ C_{1241} = C_{1443}, \ \ C_{1232} = C_{2343}, \ \ C_{1242} = C_{2434}$$

$$C_{1212} = C_{3434}, \ \ C_{1342} = \frac{1}{2}(C_{1212} - C_{1234}) = \frac{1}{2}(C_{3434} - C_{1234}) \qquad (2.22)$$

From the above discussion, the relationships between the various components of the Riemann tensor and the Weyl and the Ricci tensors, using Eq. (2.17), are given by

$$C_{1212} = R_{1212} - R_{12} + \frac{R}{6}, \ \ C_{1213} = R_{1213} - \frac{1}{2}R_{13}, \ \ C_{1223} = R_{1223} + \frac{1}{2}R_{23}$$

$$C_{1234} = R_{1234}, \ \ C_{1313} = R_{1313}, \ \ C_{1314} = 0 = R_{1314} - \frac{1}{2}R_{11} \qquad (2.23)$$

$$C_{1323} = 0 = R_{1323} + \frac{1}{2}R_{33}, \ \ C_{1324} = R_{1324} - \frac{1}{12}R_{12}, \ \ C_{1334} = R_{1334} - \frac{1}{2}R_{13}$$

$$C_{2323} = R_{2323}, \ \ C_{2334} = R_{2334} - \frac{1}{2}R_{23}, \ \ C_{3434} = R_{3434}$$

and the additional complex conjugate relations obtained by replacing index 3 by index 4 and vice versa.

In the NP formalism, the ten independent components of the Weyl tensor are represented by five (complex) scalars

$$\Psi_0 = -C_{1313} = -C_{pqrs} l^p m^q l^r m^s$$

$$\Psi_1 = -C_{1213} = -C_{pqrs} l^p n^q l^r m^s$$

$$\Psi_2 = -C_{1342} = -C_{pqrs} l^p m^q \bar{m}^r n^s \qquad (2.24)$$

$$\Psi_3 = -C_{1242} = -C_{pqrs} l^p n^q \bar{m}^r n^s$$

$$\Psi_4 = -C_{2424} = -C_{pqrs} n^p \bar{m}^q n^r \bar{m}^s$$

The tetrad components of the Weyl tensor have the following algebraic properties:
(a) If $\Psi_0 \neq 0$ and the others are zero, the gravitational field is of Petrov type N with n^i as the propagation vector;
(b) If $\Psi_1 \neq 0$ and the others are zero, the gravitational field is of Petrov type III with n^i as the propagation vector;
(c) If $\Psi_2 \neq 0$ and the others are zero, the gravitational field is of Petrov type D with l^i and n^i as the propagation vectors;
(d) If $\Psi_3 \neq 0$ and the others are zero, the gravitational field is of Petrov type III with l^i as the propagation vector;
(e) If $\Psi_4 \neq 0$ and the others are zero, the gravitational field is of Petrov type N with l^i as the propagation vector.

By a propagation vector, we mean a repeated principal null vector. If $\Psi_0 = \Psi_1 = 0$, then the gravitational field is said to be algebraically special (all Petrov types except type I), while type I gravitational field is called algebraically general.

The physical interpretation of Ψ_r $(r = 0, 1, \ldots, 4)$ is given as
Ψ_4 represents a transverse wave component in n^i direction;
Ψ_3 represents a longitudinal wave component in n^i direction;
Ψ_2 represents a Coulomb component;
Ψ_1 represents a longitudinal wave component in l^i direction;
Ψ_0 represents a transverse wave component in l^i direction.

The ten components of the Ricci tensor are given as (four real and three complex scalars)

$$\Phi_{00} = -\frac{1}{2}R_{11} = \bar{\Phi}_{00}, \quad \Phi_{22} = -\frac{1}{2}R_{22}, \quad \Phi_{02} = -\frac{1}{2}R_{33} = \bar{\Phi}_{02}$$

$$\Phi_{20} = -\frac{1}{2}R_{44}, \quad \Phi_{11} = -\frac{1}{4}(R_{12} + R_{34}), \quad \Phi_{01} = -\frac{1}{2}R_{13} = \bar{\Phi}_{10} \qquad (2.25)$$

$$\Phi_{10} = -\frac{1}{2}R_{14} = \bar{\Phi}_{01}, \quad \Phi_{12} = -\frac{1}{2}R_{23}, \quad \Phi_{21} = -\frac{1}{2}R_{24}, \quad \Lambda = \frac{1}{24}R = \frac{1}{12}(R_{12} - R_{34})$$

2.4 The Commutation Relations

Consider the commutation relation [cf., Eqs. (1.32) and (1.34)]

$$[e_{(a)}, e_{(b)}] = (\gamma_{cba} - \gamma_{cab})e^c = C_{ab}{}^c e_c \qquad (2.26)$$

Put $a = 2, b = 1$, then from Eq. (2.7), the left-hand side of Eq. (2.26) becomes

$$[e_2, e_1] = e_2 e_1 - e_1 e_2 = \Delta D - D\Delta = [\Delta, D]$$

and Eq. (2.26) can thus be expressed as

$$\Delta D - D\Delta = (\gamma_{c12} - \gamma_{c21})e^c$$

$$= (\gamma_{112} - \gamma_{121})e^1 + (\gamma_{212} - \gamma_{221})e^2 + (\gamma_{312} - \gamma_{321})e^3 + (\gamma_{412} - \gamma_{421})e^4$$

$$= -\gamma_{121}e^1 + \gamma_{212}e^2 + (\gamma_{312} - \gamma_{321})e^3 + (\gamma_{412} - \gamma_{421})e^4$$

$$= -\gamma_{121}\Delta + \gamma_{212}D + (\gamma_{312} - \gamma_{321})(-\bar{\delta}) + (\gamma_{412} - \gamma_{421})(\delta) \qquad (2.27)$$

from Eq. (2.7). Now using the definition of Ricci rotation coefficients [cf., Eqs. (1.15)] and (2.11), we get

$$\Delta D - D\Delta = (\gamma + \bar{\gamma})D + (\epsilon + \bar{\epsilon})\Delta - (\bar{\tau} + \pi)\delta - (\tau + \bar{\pi})\bar{\delta} \qquad (2.28)$$

In a similar manner, we have

$$\delta D - D\delta = (\bar{\alpha} + \beta - \bar{\pi})D + \kappa\Delta - \sigma\bar{\delta} - (\bar{\rho} + \epsilon - \bar{\epsilon})\delta \qquad (2.29)$$

$$\delta\Delta - \Delta\delta = -\bar{\nu}D + (\tau - \bar{\alpha} - \beta)\Delta + \bar{\lambda}\bar{\delta} + (\mu - \gamma + \bar{\gamma})\delta \qquad (2.30)$$

$$\bar{\delta}\delta - \delta\bar{\delta} = (\bar{\mu} - \mu)D + (\bar{\rho} - \rho)\Delta - (\bar{\alpha} - \beta)\bar{\delta} - (\bar{\beta} - \alpha)\delta \qquad (2.31)$$

2.5   The Ricci Identities (NP Field Equations)

By considering the various components of Eq. (1.36) we can write down a total of 36 equations. But in NP formalism it is sufficient to write only half the number of equations (by omitting to write down the complex conjugate of an equation). Before writing down the whole set of equations, let us work out how we can obtain such equations, for example, consider (1313) component of Eq. (1.36) [using Eqs. (2.23)–(2.25)]

$$R_{1313} = C_{1313} = -\Psi_0 = -\gamma_{131,3} + \gamma_{133,1} + \gamma_{31g}(\gamma_{13}^g - \gamma_{31}^g) + \gamma_{f11}\gamma_{33}^f - \gamma_{f13}\gamma_{31}^f$$

Now taking $f, g = 1, 2, 3, 4$ and using the properties of Ricci rotation coefficients, the above equation reduces to

$$-\Psi_0 = R_{1313} = C_{1313} = -\gamma_{131,3} + \gamma_{133,1} + \gamma_{133}(\gamma_{121} + \gamma_{431} - \gamma_{413} + \gamma_{431} + \gamma_{134})$$

$$-\gamma_{131}(\gamma_{433} + \gamma_{123} - \gamma_{213} + \gamma_{231} + \gamma_{132})$$

Now using Eqs. (2.7) and (2.11), after rearrangement of terms, the above equation leads to

$$D\sigma - \delta\kappa = \sigma(3\epsilon - \bar{\epsilon} + \rho + \bar{\rho}) + \kappa(\bar{\pi} - \tau - 3\beta - \bar{\alpha}) + \Psi_0 \qquad (2.32)$$

The complete set of 18 equations are given below, in each case we have indicated the components of the Riemann tensor which give rise to that particular equation:

$$D\rho - \bar{\delta}\kappa = (\rho^2 + \sigma\bar{\sigma}) + (\epsilon + \bar{\epsilon})\rho - \bar{\kappa}\tau - \kappa(3\alpha + \bar{\beta} - \pi) + \Phi_{00}, \ [R_{1314}]$$

(2.33a)

$$D\sigma - \delta\kappa = (\rho + \bar{\rho})\sigma + (3\epsilon - \bar{\epsilon})\sigma + (\bar{\pi} - \tau - 3\beta - \bar{\alpha})\kappa + \Psi_0, \ [R_{1313}] \quad (2.33b)$$

$$D\tau - \Delta\kappa = (\tau + \bar{\pi})\rho + (\bar{\tau} + \pi)\sigma + (\epsilon - \bar{\epsilon})\tau - (3\gamma + \bar{\gamma})\kappa + \Psi_1 + \Phi_{01}, \ [R_{1312}]$$

(2.33c)

$$D\alpha - \bar{\delta}\epsilon = (\rho + \bar{\epsilon} - 2\epsilon)\alpha + \beta\bar{\sigma} - \bar{\beta}\epsilon - \kappa\lambda - \bar{\kappa}\gamma + (\epsilon + \rho)\pi + \Phi_{01}, \ [\tfrac{1}{2}(R_{3414} - R_{1214})]$$

(2.33d)

$$D\beta - \delta\epsilon = (\alpha + \pi)\sigma + (\bar{\rho} - \bar{\epsilon})\beta - (\mu + \gamma)\gamma - (al\bar{p}ha - \bar{\pi})\epsilon + \Psi_1, \ [\tfrac{1}{2}(R_{1213} - R_{3413})]$$

(2.33e)

$$D\gamma - \Delta\epsilon = (\tau + \bar{\pi})\alpha + (\bar{\tau} + \pi)\beta - (\epsilon + \bar{\epsilon})\gamma - (\gamma + \bar{\gamma})\epsilon + \tau\pi - \nu\kappa$$

$$+ \Psi_2 - \Lambda + \Phi_{11}, \ [\tfrac{1}{2}(R_{1212} - R_{3412})] \tag{2.33f}$$

$$D\lambda - \bar{\delta}\pi = (\rho\lambda + \bar{\sigma}\mu) + \pi^2 + (\alpha - \bar{\beta})\pi - \nu\kappa - (3\epsilon - \bar{\epsilon})\lambda + \Phi_{20}, \ [R_{2441}]$$

(2.33g)

$$D\mu - \delta\pi = (\bar{\rho}\mu + \sigma\lambda) + \pi\bar{\pi} - (\epsilon + \bar{\epsilon})\mu - \pi(\bar{\alpha} - \beta) - \nu\kappa + \Psi_2 + 2\Lambda, \ [R_{2431}]$$

(2.33h)

$$D\nu - \Delta\pi = (\pi + \bar{\tau})\mu + (\bar{\pi} + \tau)\lambda + (\gamma - \bar{\gamma})\pi - (3\epsilon + \bar{\epsilon})\nu + \Psi_3 + \Phi_{21}, \ [R_{2421}]$$

(2.33i)

$$\Delta\lambda - \bar{\delta}\nu = -(\mu + \bar{\mu})\lambda - (3\gamma - \bar{\gamma})\lambda + (3\alpha + \bar{\beta} + \pi - \bar{\tau})\nu - \Psi_4, \ [R_{2442}]$$

(2.33j)

$$\delta\rho - \bar{\delta}\sigma = \rho(\bar{\alpha} + \beta) - \sigma(3\alpha - \bar{\beta}) + (\rho - \bar{\rho})\tau - (\mu - \bar{\mu})\kappa - \Psi_1 + \Phi_{01}, \ [R_{3143}]$$

(2.33k)

$$\delta\alpha - \bar{\delta}\beta = (\mu\rho - \lambda\sigma) + \alpha\bar{\alpha} + \beta\bar{\beta} - 2\alpha\beta + \gamma(\rho - \bar{\rho}) + \epsilon(\mu - \bar{\mu})$$

$$- \Psi_2 + \Lambda + \Phi_{11}, \ [\tfrac{1}{2}(R_{1234} - R_{3434})] \tag{2.33l}$$

$$\delta\lambda - \bar{\delta}\mu = -(\rho - \bar{\rho})\nu + (\mu - \bar{\mu})\pi + \mu(\alpha + \bar{\beta}) + (\bar{\alpha} - 3\beta) - \Psi_3 + \Phi_{21}, \ [R_{2443}]$$

(2.33m)

$$\delta\nu - \Delta\mu = (\mu^2 + \lambda\bar{\lambda}) + (\gamma + \bar{\gamma})\mu - \bar{\nu}\pi + (\tau - 3\beta - \bar{\alpha})\nu + \Phi_{22}, \ [R_{2423}]$$

(2.33n)

$$\delta\gamma - \Delta\beta = (\tau - \bar{\alpha} - \beta)\gamma + \mu\tau - \sigma\tau - \sigma\nu - \epsilon\bar{\nu} - \beta(\gamma - \bar{\gamma} - \mu)$$

$$+ \alpha\bar{\lambda} + \phi_{12}, \ [\tfrac{1}{2}(R_{1232} - R_{3432})] \tag{2.33o}$$

$$\delta\tau - \Delta\sigma = (\mu\sigma + \bar{\lambda}\sigma) + (\tau + \beta - \bar{\alpha})\tau - (3\gamma - \bar{\gamma})\sigma - \kappa\bar{\nu} + \Phi_{02}, \ [R_{1332}]$$

(2.33p)

$$\Delta\rho - \bar{\delta}\tau = -(\rho\bar{\mu} + \sigma\lambda) + (\bar{\beta} - \alpha - \bar{\tau})\tau + (\gamma + \bar{\gamma})\rho + \nu\kappa - \Psi_2 - \Lambda, \quad [R_{1324}]$$
$$(2.33\text{q})$$

$$\Delta\alpha - \bar{\delta}\gamma = (\rho + \epsilon)\nu - (\tau + \beta)\lambda + (\bar{\gamma} - \mu)\alpha + (\bar{\beta} - \bar{\tau})\gamma - \Psi_3, \quad [\frac{1}{2}(R_{1242} - R_{3442})]$$
$$(2.33\text{r})$$

The Bianchi identities (1.37) when written in general are very long; however, in empty spacetime ($R_{ij} = 0$), they have the following form:

$$D\Psi_0 - \bar{\delta}\Psi_0 = -3\kappa\Psi_2 + (2\epsilon + 4\rho)\Psi_1 - (-\pi + 4\alpha)\Psi_0 \qquad (2.34\text{a})$$

$$D\Psi_2 - \bar{\delta}\Psi_1 = -2\kappa\Psi_3 + 3\rho\Psi_2 - (-2\pi + 2\alpha)\Psi_4 - \lambda\Psi_0 \qquad (2.34\text{b})$$

$$D\Psi_3 - \bar{\delta}\Psi_2 = -\kappa\Psi_4 - (2\epsilon - 2\rho)\Psi_3 + 3\pi\Psi_2 - 2\lambda\Psi_1 \qquad (2.34\text{c})$$

$$D\Psi_4 - \bar{\delta}\Psi_3 = -(4\epsilon - \rho)\Psi_4 + (4\pi + 2\alpha)\Psi_3 - 3\lambda\Psi_2 \qquad (2.34\text{d})$$

$$\Delta\Psi_0 - \delta\Psi_1 = (4\gamma - \mu)\Psi_0 - (4\tau + 2\beta)\Psi_1 + 3\sigma\Psi_2 \qquad (2.34\text{e})$$

$$\Delta\Psi_1 - \delta\Psi_2 = \nu\Psi_0 + (2\gamma - 2\mu)\Psi_1 - 3\tau\Psi_2 + 2\sigma\Psi_3 \qquad (2.34\text{f})$$

$$\Delta\Psi_2 - \delta\Psi_3 = 2\nu\Psi_1 - 3\mu\Psi_2 + (2\beta - 2\tau)\Psi_3 + \sigma\Psi_4 \qquad (2.34\text{g})$$

$$\Delta\Psi_3 - \delta\Psi_4 = 3\nu\Psi_2 - (4\mu - 2\gamma)\Psi_3 + (4\beta - \tau)\Psi_4 \qquad (2.34\text{h})$$

The commutator relations [Eqs. (2.28)–(2.31)], the NP equations [Eqs. (2.33)] and the Bianchi identities [Eqs. (2.34)] form a complete set of equations from which the solution of Einstein field equations can be obtained.

Despite the fact that one has to solve a number of equations, the NP formalism has great advantages. All the differential equations are of first order. Gauge transformation of the tetrad can be used to simplify the equations. We can extract invariant properties of the gravitational field without using a coordinate basis. The algebraic structure of the Weyl tensor can be specified from the very beginning.

2.6 The Physical and Geometrical Meanings of the Spin-Coefficients and the Optical Scalars

For the physical meaning of the spin-coefficients, we consider the propagation of the basis vectors along the vectors l, n, m or $\bar{m}$. Thus, by definition, the first order change in a basis vector $e_{(a)}$ when it goes an infinitesimal displacement ξ is

$$\delta e_{(a)i} = e_{(a)i;j}\xi^j = e_i^{(b)}\gamma_{(b)(a)(c)}e_j^{(c)}\xi^j = -\gamma_{(a)(b)(c)}e_i^{(b)}\xi^{(c)} \qquad (2.35)$$

[from Eq. (1.17)]. Therefore the change, $\delta e_{(a)}^{(c)}$ in $e_{(a)}$, per unit displacement along the direction c, is

$$\delta e_{(a)}^{(c)} = -\gamma_{(a)(b)(c)}e^{(b)} \tag{2.36}$$

So, in particular, the change in l, per unit displacement along l [using Eqs. (2.6), (2.36) and (2.11)] is

$$\delta l(l) = -\gamma_{1(b)1}e^{(b)} = -\gamma_{111}e^1 - \gamma_{121}e^2 - \gamma_{131}e^3 - \gamma_{141}e^4$$

$$= -\gamma_{121}l + \gamma_{131}\bar{m} + \gamma_{141}m = (\epsilon + \bar{\epsilon})l - \kappa\bar{m} + \bar{\kappa}m \tag{2.37}$$

and

$$\delta n(l) = -\gamma_{2(b)1}e^{(b)} = -(\epsilon + \bar{\epsilon})n + \pi m + \bar{\pi}\bar{m} \tag{2.38}$$

$$\delta m(l) = -\gamma_{3(b)1}e^{(b)} = -(\epsilon - \bar{\epsilon})m + \bar{\pi}l - \kappa n \tag{2.39}$$

Equation (2.37) can also be written as

$$l_{i;j}l^j = (\epsilon + \bar{\epsilon})l_i - \kappa\bar{m}_i + \bar{\kappa}m_i \tag{2.40}$$

We also have

$$l_{i;j} = \gamma_{(a)1(b)}e_i^{(a)}e_j^{(b)}$$

$$= (\gamma + \bar{\gamma})l_jl_i - \bar{\tau}l_jm_i - \tau l_j\bar{m}_i + (\epsilon + \bar{\epsilon})n_jl_i - \bar{\kappa}n_jm_i - \kappa n_j\bar{m}_i$$

$$- (\alpha + \bar{\beta})m_jl_i + \bar{\sigma}m_jm_i + \rho m_j\bar{m}_i - (\bar{\alpha} + \beta)\bar{m}_jl_i + \bar{\rho}\bar{m}_jm_i + \sigma\bar{m}_j\bar{m}_i \tag{2.41}$$

[using Eq. (2.11)]. It may be noted that contracting Eq. (2.41) with l^j, we get Eq. (2.40).

In a similar manner, we can obtain

$$n_{i;j} = -(\gamma + \bar{\gamma})l_jn_i - \nu l_jm_i + \bar{\nu}l_j\bar{m}_i - (\epsilon + \bar{\epsilon})n_jn_i + \pi n_jm_i + \bar{\pi}n_j\bar{m}_i$$

$$+ (\alpha + \bar{\beta})m_jn_i - \lambda m_jm_i - \bar{\mu}m_j\bar{m}_i + (\bar{\alpha} + \beta)\bar{m}_jn_i - \mu\bar{m}_jm_i - \bar{\lambda}\bar{m}_j\bar{m}_i \tag{2.42}$$

$$m_{i;j} = \bar{\nu}l_jl_i - \tau l_jn_i + (\gamma - \bar{\gamma})l_jm_i + \bar{\pi}n_jl_i - \kappa n_jn_i + (\epsilon - \bar{\epsilon})n_jm_i - \bar{\mu}m_jl_i$$

$$+ \rho m_jn_i + (\bar{\beta} - \alpha)m_j\bar{m}_i - \bar{\lambda}\bar{m}_jl_i + \sigma\bar{m}_jn_i + (\bar{\alpha} - \beta)\bar{m}_jm_i \tag{2.43}$$

The contraction of these equations leads to

$$l^i{}_{;i} = (\epsilon + \bar{\epsilon}) - (\rho + \bar{\rho}) \tag{2.44}$$

$$n^i{}_{;i} = (\mu + \bar{\mu}) - (\gamma + \bar{\gamma}) \tag{2.45}$$

$$m^i{}_{;i} = -\bar{\alpha} + \bar{\pi} - \tau + \beta \tag{2.46}$$

The geometrical meanings of different spin-coefficients can be summarized in the following theorems (for the proofs of these theorems, the reader is referred to [9]).

Theorem 2.1 *The null congruence $\Gamma(l)$ is geodesic if and only if $\kappa = 0$ and by an appropriate choice of affine parameter along $\Gamma(l)$ we may choose $\epsilon + \bar{\epsilon} = 0$ (or, $\epsilon = 0$).*

Theorem 2.2 *The null congruence $\Gamma(n)$ is geodesic if and only if $\nu = 0$ and by an appropriate choice of affine parameter along $\Gamma(n)$ we may choose $\gamma + \bar{\gamma} = 0$ (or, $\gamma = 0$).*

Theorem 2.3 *If we choose the scaling $\epsilon + \bar{\epsilon} = 0$ then the tetrad $\{l^i, n^i, m^i, \bar{m}^i\}$ is parallelly propagated along $\Gamma(l)$ when $\kappa = \pi = \epsilon = 0$.*

Theorem 2.4 *If we choose the scaling $\gamma + \bar{\gamma} = 0$ then the tetrad $\{l^i, n^i, m^i, \bar{m}^i\}$ is parallelly propagated along $\Gamma(n)$ when $\nu = \tau = \gamma = 0$.*

If $\kappa = \epsilon = 0$, then Eq. (2.41) reduces to

$$l_{i;j} = (\gamma + \bar{\gamma})l_j l_i - \bar{\tau}l_j m_i - \tau l_j \bar{m}_i - (\alpha + \bar{\beta})m_j l_i + \bar{\sigma}m_j m_i + \rho m_j \bar{m}_i$$

$$- (\bar{\alpha} + \beta)\bar{m}_j l_i + \bar{\rho}\bar{m}_j m_i + \sigma \bar{m}_j \bar{m}_i \tag{2.47}$$

which leads to

$$l_{[i;j]} = -(\bar{\alpha} + \beta - \tau)l_{[i}\bar{m}_{j]} - (\alpha + \bar{\beta} - \bar{\tau})l_{[i}m_{j]} - (\rho + \bar{\rho})\bar{m}_{[i}m_{j]} \tag{2.48}$$

and

$$l_{[i;j}l_{k]} = (\rho - \bar{\rho})\bar{m}_{[i}m_j l_{k]} \tag{2.49}$$

We also have the following theorems.

Theorem 2.5 *The null congruence $\Gamma(l)$ is geodesics if and only if $l^{[i}Dl^{j]} = 0$.*

Theorem 2.6 *The null congruence $\Gamma(n)$ is geodesics if and only if $n^{[i}\Delta n^{j]} = 0$.*

From Eq. (2.49), we have the following theorems.

Theorem 2.7 *Let $\Gamma(l)$ be a null geodesics congruence then $l_{[i;j}l_{k]} = 0$ is equivalent to $\rho = \bar{\rho}$.*

Theorem 2.8 *Let $\Gamma(n)$ be a null geodesics congruence then $n_{[i;j}n_{k]} = 0$ is equivalent to $\mu = \bar{\mu}$.*

Since $l_{[i;j}l_{k]} = 0$ implies that l^i is hypersurface orthogonal (i.e. l^i is proportional to the gradient of a scalar field), we have the following theorems.

Theorem 2.9 *The null vector field l is hypersurface orthogonal if $\kappa = 0$ and $\rho = \bar{\rho}$.*

Theorem 2.10 *The null vector field n is hypersurface orthogonal if $\nu = 0$ and $\mu = \bar{\mu}$.*

Thus, we can say that the congruence of null geodesic is hypersurface orthogonal if and only if ρ is real, and l^i will be equal to the gradient of a scalar field if and only if in addition $\tau = \bar{\alpha} + \beta$.

From Eqs. (2.47)–(2.49), we have

$$\frac{1}{2}l^i{}_{;i} = -\frac{1}{2}(\rho + \bar{\rho}) = \theta \tag{2.50}$$

$$\frac{1}{2}l_{[i;j]}l^{i;j} = -\frac{1}{4}(\rho - \bar{\rho})^2 = \omega^2 \tag{2.51}$$

$$\frac{1}{2}l_{(i;j)}l^{i;j} = \theta^2 + \mid \sigma \mid^2 \tag{2.52}$$

The quantities θ, ω and σ are, respectively, the expansion, the twist and the shear of the congruence, and all of them are called the *optical scalars*. θ and ω are also defined as

$$\theta = -\mathrm{Re}\,\rho, \quad \omega = \mathrm{Im}\,\rho \tag{2.53}$$

If we take $\kappa = \epsilon = 0$, then the equations describing the behaviour of ρ and σ along the geodesic are, respectively, given by the Eqs. (2.33a) and (2.33b) and we have

$$D\rho = \rho^2 + \sigma\bar{\sigma} + \Phi_{00} = \rho^2 + \mid \sigma \mid^2 + \Phi_{00} \tag{2.54}$$

$$D\sigma = \sigma(\rho + \bar{\rho}) + \Psi_0 \tag{2.55}$$

where

$$\Phi_{00} = -\frac{1}{2}R_{11} = -\frac{1}{2}R_{ij}l^i l^j, \quad \Psi_0 = -C_{pqrs}l^p m^q l^r m^s \tag{2.56}$$

From Eq. (2.50), Eq. (2.55) can be written as

$$D\sigma = -2\theta\sigma + \Psi_0 \tag{2.57}$$

Taking the imaginary part of Eq. (2.54) (as Φ_{00} is real), we have

$$D\omega = \frac{1}{2}(\rho + \bar{\rho})(\rho - \bar{\rho}) = -2\theta\omega \tag{2.58}$$

while taking the real part of Eq. (2.54), we get

$$D\theta = \omega^2 - \theta^2 - \mid \sigma \mid^2 - \Phi_{00} \tag{2.59}$$

Equations (2.57)–(2.59) are the standard equations representing the variation of shear, rotation (twist) and expansion, respectively.

Finally, we state the following theorem.

Goldberg–Sachs Theorem. If $\kappa = \sigma = 0$ then the gravitational field is algebraically special and conversely.

References

1. Ahsan, Z.: Proc. Ind. Acad. Sc. **88A**, 125–132 (1979)
2. Ahsan, Z.: J. Moscow Phys. Soc. **5**, 215–222 (1995)
3. Ahsan, Z.: Acta Phys. Sin. **4**, 337–343 (1995)
4. Ahsan, Z.: Indian J. Pure Appl. Maths. **31**(2), 215–225 (2000)
5. Ahsan, Z.: Tensors: Mathematics of Differential Geometry and Relativity, Second Printing. PHI Learning Pvt. Ltd, Delhi (2017)
6. Ahsan, Z., Ali, M.: Int. J. Theo. Phys. **51**, 2044–2055 (2012)
7. Ahsan, Z., Ali, M.: Int. J. Theo. Phys. **54**, 1397–1407 (2015)
8. Chandrashekhar, S.: Mathematical Theory of Black Holes. Oxford University Press, New York (1983)
9. Frolov, V.P.: The Newman-Penrose method in theory of general relativity. In: Basov, N.G. (ed.) Problems in General Theory of Relativity and the Theory of Group Representation. Plenum Press, New York (1979)
10. Lind, R.W.: Gen. Rel. Grav. **5**, 25 (1974)
11. Lind, R.W., Messmer, J., Newman, E.T.: J. Math. Phys. **13**, 1884 (1972)
12. Ludwig, G.: Am. J. Phys. **37**, 1225 (1969)
13. Newman, E.T., Penrose, R.: J. Math. Phys. **3**, 566–578 (1962)
14. Penrose, R.: Ann. Phys. **10**, 171 (1960)
15. Penrose, R., Rindler, W.: Spinors and Spacetime. Cambridge University Press, New York (1984)
16. Petrov, A.Z.: Sc. Not. Kazan State Univ. **114**, 55 (1954)
17. Petrov, A.Z.: Dissertation, Moscow State University (1957)
18. Pirani, F.A.E., Trautman, A., Bondi, H.: Lectures on General Relativity: Brandeis Summer Institute in Theoretical Physics. Prentice-Hall, N. J. (1965)
19. Sachs, R.K.: Proc. Roy. Soc. Lond. A **264**, 309 (1961)
20. Stephani, H., Kramer, D., Maccallum, M., Hoenselaers, C., Herlt, E.: Exact Solutions of Einstein's Field Equations, 2nd edn. Cambridge University Press, UK (2003)
21. Synge, J.: Comm. Dublin Inst. Adv. St. A **15** (1969)
22. Talbot, C.J.: J. Math. Phys. **13**, 45 (1969)

Chapter 3
The Geroch–Held–Penrose Formalism

3.1 Introduction

In the application of the tetrad formalism, the choice of the tetrad basis depends upon the symmetries of the spacetime with which we are working, and to some extent is the part of the problem. In the previous chapter, we have discussed an important example of the tetrad formalism—NP formalism [6]. This is a tetrad formalism with special choice of null basis consisting of a pair of real null vectors l^a and n^a and a pair of complex conjugate null vectors m^a and $\bar{m}^a$. These vectors satisfy the orthogonality conditions

$$l_a m^a = l_a \bar{m}^a = n_a m^a = n_a \bar{m}^a = 0 \tag{3.1}$$

besides the requirement

$$l_a l^a = n_a n^a = m_a m^a = \bar{m}_a \bar{m}^a = 0 \tag{3.2}$$

that the vectors be null. The basis vectors also satisfy the normalization condition

$$l_a n^a = 1 \ , \quad m_a \bar{m}^a = -1 \tag{3.3}$$

Since the chosen tetrad is null, it is not surprising that this formalism has an alternative and more general definitions in terms of spinors [7].

Although such formalisms are useful in general, they have particular advantage if the basis vectors (or, spinors) are not completely arbitrary but are related to the geometry or physics in some natural way. For example, if we are studying the geometry of a null 2-surface S, we can then choose the tetrad so that l^a and n^a point along the outgoing and ingoing normal to S, and the real and imaginary parts of m^a and $\bar{m}^a$ are tangent to S. The remaining gauge freedom in the choice of the tetrad is the two-dimensional subgroup of the Lorentz group representing a boost in the direction normal to S and a rotation in the direction tangent to S. In terms of the spinors, we

© Springer Nature Singapore Pte Ltd. 2019
Z. Ahsan, *The Potential of Fields in Einstein's Theory of Gravitation*,
https://doi.org/10.1007/978-981-13-8976-4_3

choose the flagpoles of o^A and ι^A to point along the directions of the null normals and the remaining gauge freedom (which preserves the normalization $o_A \iota^A = 1$) is

$$o^A \longrightarrow \lambda o^A \ , \ \iota^A \longrightarrow \lambda^{-1} \iota^A \tag{3.4}$$

where λ is an arbitrary (nowhere vanishing) complex scalar field. The pair of spinor fields o^A and ι^A is called a *dyad* or *spin frame*. Other important physical situations in which the tetrad is defined up to spin and boost transformations include certain radiation problems and the case of Petrov type D spacetime.

Under the transformation (3.4), some of the spin-coefficients (Ricci rotation coefficients with respect to a basis) are simply rescaled, while the other transform in a way which includes the derivative of λ. It turns out that the spin-coefficients can be combined with the differential operators to produce new differential operators of proper spin and boost. If we deal with the quantities that simply rescale under spin and boost transformations, we then have Geroch–Held–Penrose formalism (or, in short, GHP formalism) or, compacted spin-coefficient formalism [3]. This formalism is more concise and efficient than the widely known NP formalism; however, the GHP formalism has failed to develop its full potential to the extent to which the NP formalism has. Soon after the appearance of GHP formalism, Held [4, 5] proposed a simple procedure for integration within this formalism and applied it successfully to Petrov type D vacuum metrics.

The present chapter is devoted to the study of this formalism. The salient features of the formalism are discussed in Sect. 3.2; while in Sect. 3.3, a complete set of field equations, commutator relations and Bianchi identities has been given. In the last section, the geometric and physical meanings of null congruences are discussed along with the related results (a detailed account of this formalism and its applications along with the Newman–Penrose formalism have been given by Ahsan [1]).

3.2 Spacetime Calculus

The most general transformation, in terms of spinor notation, preserving the two preferred null directions and the dyad normalization $o_A \iota^A$ is given by Eq. (3.4). The corresponding two parameter subgroup of the Lorentz group (boost and spatial rotation) affects the complex null tetrad as follows:

$$l^a \longrightarrow r l^a \ , \ n^a \longrightarrow r^{-1} n^a \ \text{(boost)} \tag{3.5}$$

$$m^a \longrightarrow e^{i\theta} m^a \ \ \text{(spatial rotation)} \tag{3.6}$$

where the complex vector m^a is defined by $m^a = \frac{1}{\sqrt{2}}(X^a + iY^a)$, X^a and Y^a are the unit space-like vectors orthogonal to each of l^a, n^a and to each other, and r and θ are related through $\lambda^2 = re^{i\theta}$. In terms of the tetrad, the transformation (3.4) takes the form

$$l^a \longrightarrow \lambda \bar{\lambda} l^a \ , \ n^a \longrightarrow \lambda^{-1} \bar{\lambda}^{-1} n^a \ , \ m^a \longrightarrow \lambda \bar{\lambda}^{-1} m^a \ , \ \bar{m}^a \longrightarrow \lambda^{-1} \bar{\lambda} \bar{m}^a \quad (3.7)$$

The GHP formalism deals with the scalars associated with a tetrad $\{l^a, n^a, m^a, \bar{m}^a\}$ /dyad (o^A, ι^A) where the scalars undergo transformation

$$\eta \longrightarrow \lambda^p \bar{\lambda}^q \eta \tag{3.8}$$

whenever the tetrad/dyad is changed according to Eqs. (3.5) and (3.6) or Eqs. (3.7)/(3.4). Such a scalar is called a *spin* and *boost weighted scalar of type* $\{p, q\}$. The *spin weight* is $\frac{1}{2}(p - q)$ and the *boost weight* is $\frac{1}{2}(p + q)$. It may be noted that o^A and ι^A may themselves be regarded as spinors of type $\{1, 0\}$ and $\{-1, 0\}$, respectively, and $l^a, n^a, m^a, \bar{m}^a$ as vectors of types $\{1, 1\}, \{-1, -1\}, \{1, -1\}, \{-1, 1\}$, respectively.

It is known that any dyad defines a unique null tetrad $Z_\mu^a = \{l^a, n^a, m^a, \bar{m}^a\}$ at each point and, conversely, that any null tetrad defines a dyad uniquely up to sign. The relationship is as follows:

$$l^a = o^A \bar{o}^{A'} \ , \ n^a = \iota^A \bar{\iota}^{A'} \ , \ m^a = o^A \bar{\iota}^{A'} \ , \ \bar{m}^a = \iota^A \bar{o}^{A'} \tag{3.9}$$

The 12 spin-coefficients (complex functions) are as follows:

$$\kappa = o^A \bar{o}^{A'} o^B \nabla_{AA'} o_B = m^b l^a \nabla_a l_b$$

$$\sigma = o^A \bar{\iota}^{A'} o^B \nabla_{AA'} o_B = m^b m^a \nabla_a l_b \tag{3.10a}$$

$$\rho = \iota^A \bar{o}^{A'} o^B \nabla_{AA'} o_B = m^b \bar{m}^a \nabla_a l_b$$

$$\tau = \iota^A \bar{\iota}^{A'} o^B \nabla_{AA'} o_B = m^b n^a \nabla_a l_b$$

$$\kappa' = - \iota^A \bar{\iota}^{A'} \iota^B \nabla_{AA'} \iota_B = \bar{m}^b n^a \nabla_a n_b$$

$$\sigma' = - \iota^A \bar{o}^{A'} \iota^B \nabla_{AA'} \iota_B = \bar{m}^b \bar{m}^a \nabla_a n_b \tag{3.10b}$$

$$\rho' = - o^A \bar{\iota}^{A'} \iota^B \nabla_{AA'} \iota_B = \bar{m}^b m^a \nabla_a n_b$$

$$\tau' = - o^A \bar{o}^{A'} \iota^B \nabla_{AA'} \iota_B = \bar{m}^b l^a \nabla_a n_b$$

$$\beta = o^A \bar{\iota}^{A'} \iota^B \nabla_{AA'} o_B = \frac{1}{2} (n^b m^a \nabla_a l_b - \bar{m}^b m^a \nabla_a m_b)$$

$$\beta' = - \iota^A \bar{o}^{A'} o^B \nabla_{AA'} \iota_B = \frac{1}{2} (l^b \bar{m}^a \nabla_a n_b - m^b \bar{m}^a \nabla_a \bar{m}_b) \tag{3.10c}$$

$$\epsilon = o^A \bar{o}^{A'} \iota^B \nabla_{AA'} o_B = \frac{1}{2} (n^b l^a \nabla_a l_b - \bar{m}^b l^a \nabla_a m_b)$$

$$\epsilon' = -\iota^A \bar{\iota}^{A'} o^B \nabla_{AA'} \iota_B = \frac{1}{2}(l^b n^a \nabla_a n_b - m^b n^a \nabla_a \bar{m}_b)$$

We recall the definitions of the tetrad (dyad) components of Weyl tensor C_{abcd} (Weyl spinor Ψ_{ABCD}) and the trace-free Ricci tensor R_{ab} (Ricci spinor $\Phi_{ABC'D'}$):

$$\Psi_0 = -C_{abcd} l^a m^b l^c m^d = o^A o^B o^C o^D \Psi_{ABCD} = \Psi_4'$$

$$\Psi_1 = -C_{abcd} l^a n^b l^c m^d = o^A o^B o^C \iota^D \Psi_{ABCD} = \Psi_3'$$

$$\Psi_2 = -\frac{1}{2} C_{abcd}(l^a n^b l^c n^d + l^a n^b m^c \bar{m}^d) = o^A o^B \iota^C \iota^D \Psi_{ABCD} = \Psi_2' \qquad (3.11)$$

$$\Psi_3 = -C_{abcd} l^a n^b n^c \bar{m}^d = o^A \iota^B \iota^C \iota^D \Psi_{ABCD} = \Psi_1'$$

$$\Psi_4 = -C_{abcd} n^a \bar{m}^b n^c \bar{m}^d = \iota^A \iota^B \iota^C \iota^D \Psi_{ABCD} = \Psi_0'$$

$$\Phi_{00} = -\frac{1}{2} R_{11} = o^A o^B \bar{o}^{A'} \bar{o}^{B'} \Phi_{ABA'B'} = \bar{\Phi}_{00} = \Phi_{22}'$$

$$\Phi_{01} = -\frac{1}{2} R_{13} = o^A o^B \bar{o}^{A'} \bar{\iota}^{B'} \Phi_{ABA'B'} = \bar{\Phi}_{10} = \Phi_{21}'$$

$$\Phi_{02} = -\frac{1}{2} R_{33} = o^A o^B \bar{\iota}^{A'} \bar{\iota}^{B'} \Phi_{ABA'B'} = \bar{\Phi}_{20} = \Phi_{20}'$$

$$\Phi_{10} = -\frac{1}{2} R_{14} = o^A \iota^B \bar{o}^{A'} \bar{o}^{B'} \Phi_{ABA'B'} = \bar{\Phi}_{01} = \Phi_{12}' \qquad (3.12)$$

$$\Phi_{11} = -\frac{1}{4}(R_{12} + R_{34}) = o^A \iota^B \bar{o}^{A'} \bar{\iota}^{B'} \Phi_{ABA'B'} = \bar{\Phi}_{11} = \Phi_{11}'$$

$$\Phi_{12} = -\frac{1}{2} R_{23} = o^A \iota^B \bar{\iota}^{A'} \bar{\iota}^{B'} \Phi_{ABA'B'} = \bar{\Phi}_{21} = \Phi_{10}'$$

$$\Phi_{20} = -\frac{1}{2} R_{44} = \iota^A \iota^B \bar{o}^{A'} \bar{o}^{B'} \Phi_{ABA'B'} = \bar{\Phi}_{02} = \Phi_{02}'$$

$$\Phi_{21} = -\frac{1}{2} R_{24} = \iota^A \iota^B \bar{o}^{A'} \bar{\iota}^{B'} \Phi_{ABA'B'} = \bar{\Phi}_{12} = \Phi_{01}'$$

$$\Phi_{22} = -\frac{1}{2} R_{22} = \iota^A \iota^B \bar{\iota}^{A'} \bar{\iota}^{B'} \Phi_{ABA'B'} = \bar{\Phi}_{22} = \Phi_{00}'$$

The scalar curvature is defined by

$$\Lambda = \bar{\Lambda} = \Lambda' = \frac{1}{24}R \tag{3.13}$$

We shall also make use of the prime systematically to denote the operation of effecting the replacement

$$o^A \longrightarrow i\iota^A , \ \iota^A \longrightarrow io^A , \ \bar{o}^{A'} \longrightarrow i\bar{\iota}^{A'} , \ \bar{\iota}^A \longrightarrow -i\bar{o}^{A'} \tag{3.14}$$

so that

$$(l^a)' = n^a , \ (n^a)' = l^a , \ (m^a)' = \bar{m}^a , \ (\bar{m}^a)' = m^a \tag{3.15}$$

This preserves normalization $o_A \iota^A = 1$ and the relationship between a quantity and its complex conjugate. Since the bar and prime commute, we can write $\bar{\eta}'$ without ambiguity. Moreover, the prime operation is involutory up to sign

$$(\eta')' = (-1)^{p+q}\eta \tag{3.16}$$

(for all quantities explicitly defined in this chapter, $p + q$ is in fact even and thus this sign will play no role here).

The components of the Weyl tensor, the Ricci tensor and the spin-coefficients have the spin and boost of the types as indicated below:

$$\Psi_0 : \{4, 0\} , \ \Psi_1 : \{2, 0\} , \ \Psi_2 : \{0, 0\} , \ \Psi_3 : \{-2, 0\} , \ \Psi_4 : \{-4, 0\}$$

$$\Phi_{00} : \{2, 2\} , \ \Phi_{01} : \{2, 0\} , \ \Phi_{02} : \{2, -2\} , \ \Phi_{10} : \{0, 2\} , \ \Phi_{11} : \{2, 2\}$$

$$\Phi_{22} : \{-2, -2\} , \ \Phi_{21} : \{-2, 0\} , \ \Phi_{20} : \{-2, 2\} , \ \Phi_{12} : \{0, -2\}$$

$$\Lambda = \bar{\Lambda} = \Lambda' = \frac{1}{24}R : \{0, 0\} \tag{3.17}$$

$$\kappa : \{3, 1\} , \ \sigma : \{3, -1\} , \ \rho : \{1, 1\} , \ \tau : \{1, -1\}$$

$$\kappa' : \{-3, -1\} , \ \sigma' : \{-3, 1\} , \ \rho' : \{-1, -1\} , \ \tau' : \{-1, 1\}$$

where the spin-coefficients κ', σ', etc., defined in Eq. (3.10) are related to the spin-coefficients defined by Newman and Penrose [cf., Eq. (2.11)] as follows:

$$\nu = -\kappa', \lambda = -\sigma', \mu = -\rho', \pi = -\tau', \alpha = -\beta', \gamma = -\epsilon' \tag{3.18}$$

Out of the 12 spin-coefficients [cf., Eq. (3.10)] only eight, given by Eqs. (3.10a) and (3.10b), are of good spin and boost; and the remaining four, as defined by Eq. (3.10c), appear in the definition of the derivatives so that the derivative may not behave badly under spin and boost transformations. For such a scalar η of type $\{p, q\}$, the derivative operators are defined as

$$\mathcal{P}\eta = (D - p\epsilon - q\bar{\epsilon})\eta , \ \mathcal{P}'\eta = (D' + p\epsilon' + q\bar{\epsilon}')\eta$$

$$\mathcal{D}\eta = (\delta - p\beta + q\bar{\beta}')\eta , \ \mathcal{D}'\eta = (\delta' + p\beta' - q\bar{\beta})\eta \qquad (3.19)$$

where the symbols $\mathcal{P}$ and $\mathcal{D}$ are pronounced as *thorn* and *e(d)th*; $\mathcal{P}$ and $\mathcal{D}$ are the phonetic symbols for the soft and hard '*th*', respectively, and the types of these derivatives are

$$\mathcal{P} : \{1, 1\} , \ \mathcal{P}' : \{-1, -1\} , \ \mathcal{D} : \{1, -1\} , \ \mathcal{D}' : \{-1, 1\}$$

Alternatively, the operators may be defined in terms of the type $\{0, 0\}$ operator (acting on a quantity of type $\{p, q\} = \{r + s, r - s\}$)

$$\Theta_{AA'} = \nabla_{AA'} - p\iota^B \nabla_{AA'} o^B - q\bar{\iota}^{B'} \nabla_{AA'} \bar{o}^{B'} = \nabla_a - rn^b \nabla_a l_b + sm^b \nabla_a m_b$$
$$(3.19a)$$

by the equation

$$\Theta_a = l_a \mathcal{P}' + n_a \mathcal{P} - m_a \mathcal{D}' - \bar{m}_a \mathcal{D} \qquad (3.19b)$$

In Eq. (3.19a), s and r are the spin and boost weights, respectively. The original definition (3.19) can be obtained by transvecting Eq. (3.19b) with l^a, n^a, m^a and $\bar{m}^a$.

The basic quantities with which we are concerned here are the eight spin-coefficients $\kappa, \sigma, \rho, \tau$; $\kappa', \sigma', \rho', \tau'$ and the four differential operators $\mathcal{P}, \mathcal{D}, \mathcal{P}', \mathcal{D}'$. There is the operation of complex conjugation and also we may consider the prime as effectively an allowable operation on the system.

Also, we define

$$\bar{\mathcal{P}} = \mathcal{P} , \ \bar{\mathcal{P}}' = \mathcal{P}' , \ \bar{\mathcal{D}} = \mathcal{D}' , \ \bar{\mathcal{D}}' = \mathcal{D} \qquad (3.20)$$

then the operation of complex conjugation will satisfy

$$\overline{\mathcal{P}\eta} = \bar{\mathcal{P}}\bar{\eta} , \ \overline{\mathcal{D}\eta} = \bar{\mathcal{D}}\bar{\eta} \qquad (3.21)$$

Also, if we prime an element of type $\{p, q\}$, we get an element of type $\{-p, -q\}$. The prime will commute with addition, multiplication and the complex conjugate [but not Eq. (3.16)]. Moreover, we have

$$(\mathcal{P}\eta)' = \mathcal{P}'\eta' , \ (\mathcal{P}'\eta)' = \mathcal{P}\eta' , \ (\mathcal{D}\eta)' = \mathcal{D}'\eta' , \ (\mathcal{D}'\eta)' = \mathcal{D}\eta' \qquad (3.22)$$

3.3 GHP Equations

As a consequence of the above considerations, the Newman–Penrose equation (2.33), the commutator relations (2.28)–(2.31) and the Bianchi identities (2.34) get new explicit forms. They contain scalars and derivative operators of good spin and boost weights only and split into two sets of equations - one being the primed version of the other.

Since the complete set of these equations is not available in literature, we shall present them here.

GHP Field Equations

$$\mathcal{P}\rho - \mathcal{D}\kappa = \rho^2 + \sigma\bar{\sigma} - \bar{\kappa}\tau - \tau'\kappa + \Phi_{00} \tag{3.23a}$$

$$\mathcal{P}'\rho' - \mathcal{D}'\kappa' = \rho'^2 + \sigma'\bar{\sigma}' - \bar{\kappa}'\tau' - \tau\kappa' + \Phi_{22} \tag{3.23a'}$$

$$\mathcal{P}\sigma - \mathcal{D}\kappa = (\rho + \bar{\rho})\sigma - (\tau + \bar{\tau}')\kappa + \Psi_0 \tag{3.23b}$$

$$\mathcal{P}'\sigma' - \mathcal{D}'\kappa' = (\rho' + \bar{\rho}')\sigma' - (\tau' + \bar{\tau})\kappa' + \Psi_4 \tag{3.23b'}$$

$$\mathcal{P}\tau - \mathcal{P}'\kappa = (\tau - \bar{\tau}')\rho + (\bar{\tau} - \tau')\sigma + \Psi_1 + \Phi_{01} \tag{3.23c}$$

$$\mathcal{P}'\tau' - \mathcal{P}\kappa' = (\tau' - \bar{\tau})\rho' + (\bar{\tau}' - \tau)\sigma' + \Psi_3 + \Phi_{21} \tag{3.23c'}$$

$$\mathcal{D}\rho - \mathcal{D}'\sigma = (\rho - \bar{\rho})\tau + (\rho' - \bar{\rho}')\kappa + \Psi_1 + \Phi_{01} \tag{3.23d}$$

$$\mathcal{D}'\rho' - \mathcal{D}\sigma' = (\rho' - \bar{\rho}')\tau' + (\rho - \bar{\rho})\kappa' + \Psi_3 + \Phi_{21} \tag{3.23d'}$$

$$\mathcal{D}\tau - \mathcal{P}'\sigma = -\rho'\sigma - \bar{\sigma}'\rho + \tau^2 + \kappa\bar{\kappa}' + \Phi_{02} \tag{3.23e}$$

$$\mathcal{D}'\tau' - \mathcal{P}\sigma' = -\rho\sigma' - \bar{\sigma}\rho' + \tau'^2 + \kappa'\bar{\kappa} + \Phi_{20} \tag{3.23e'}$$

$$\mathcal{P}'\rho - \mathcal{D}'\tau = -\rho\bar{\rho}' - \tau\bar{\tau} - \kappa\kappa' - \Psi_2 - 2\Lambda \tag{3.23f}$$

$$\mathcal{P}\rho' - \mathcal{D}\tau' = -\rho'\bar{\rho} - \tau'\bar{\tau}' - \kappa'\kappa - \Psi_2 - 2\Lambda \tag{3.23f'}$$

The above list does not completely exhaust all NP field equations. The remaining equations refer to the derivatives of the spin-coefficients which are spin and boost weighted quantities, and therefore cannot be written like above equations. Instead, they play their role as part of the commutator equations for the differential operators $\mathcal{P}, \mathcal{P}', \mathcal{D}, \mathcal{D}'$. These commutators when applied to a spin and boost weighted quantity η of type $\{p, q\}$, are given as follows.

GHP Commutator Relations

$$[\mathcal{P}, \mathcal{P}']\eta = \{(\bar{\tau} - \tau')\mathcal{D} + (\tau - \bar{\tau}')\mathcal{D}' - p(\kappa\kappa' - \tau\tau' + \Psi_2 + \Phi_{11} - \Lambda)$$
$$- q(\bar{\kappa}\bar{\kappa}' - \bar{\tau}\bar{\tau}' + \Psi_2 + \Phi_{11} - \Lambda)\}\eta \tag{3.24a}$$

$$[\mathcal{P}, \mathcal{D}]\eta = \{\bar{\rho}\mathcal{D} + \sigma\mathcal{D}' - \bar{\tau}'\mathcal{P} - \kappa\mathcal{P}' - p(\rho'\kappa - \tau'\sigma + \Psi_1)$$
$$- q(\bar{\sigma}'\bar{\kappa} - \bar{\rho}\bar{\tau}' + \Phi_{01})\}\eta \tag{3.24b}$$

$$[\mathcal{D}, \mathcal{D}']\eta = \{(\bar{\rho}' - \rho')\mathcal{P} + (\rho - \bar{\rho}')\mathcal{P}' - p(\rho\rho' - \sigma\sigma' + \Psi_2 - \Phi_{11} - \Lambda)$$
$$- q(\bar{\rho}\bar{\rho}' - \bar{\sigma}\bar{\sigma}' + \Psi_2 - \Phi_{11} - \Lambda)\}\eta \tag{3.24c}$$

together with the remaining commutator relations obtained by applying prime, complex conjugation, and both to Eq. (3.24b). Care must be taken when applying primes and bars to these equations, as η', $\bar{\eta}$ and $\bar{\eta}'$ have types different to that of η. Under the prime, p becomes $-p$ and q becomes $-q$; under bar, p becomes q and q becomes p; under both bar and prime, p becomes $-q$ and q becomes $-p$.

GHP Bianchi Identities (Full)

$$\mathcal{P}\Psi_1 - \mathcal{D}'\Psi_0 - \mathcal{P}\Phi_{01} + \mathcal{D}\Phi_{00} = -\tau'\Psi_0 + 4\rho\Psi_1 - 3\kappa\Psi_2$$
$$+ \bar{\tau}'\Phi_{00} - 2\bar{\rho}\Phi_{01} - 2\sigma\Phi_{10} + 2\kappa\Phi_{11} - \bar{\kappa}\Phi_{02} \tag{3.25a}$$

$$\mathcal{P}'\Psi_3 - \mathcal{D}\Psi_4 - \mathcal{P}'\Phi_{21} + \mathcal{D}'\Phi_{22} = -\tau\Psi_4 + 4\rho'\Psi_3 - 3\kappa'\Psi_2 \tag{3.25a'}$$
$$+ \bar{\tau}\Phi_{22} - 2\bar{\rho}'\Phi_{21} - 2\sigma'\Phi_{12} + 2\kappa'\Phi_{11} - \bar{\kappa}'\Phi_{20}$$

$$\mathcal{P}\Psi_2 - \mathcal{D}'\Psi_1 - \mathcal{D}'\Phi_{01} + \mathcal{P}'\Phi_{00} + 2\mathcal{P}\Lambda = \sigma'\Psi_0 - 2\tau\Psi_1 + 3\rho\Psi_2$$
$$- 3\kappa\Psi_3 - \bar{\rho}'\Phi_{00} - 2\bar{\tau}\Phi_{10} - 2\tau\Phi_{10} + 2\rho\Phi_{11} + \bar{\sigma}\Phi_{02} \tag{3.25b}$$

$$\mathcal{P}'\Psi_2 - \mathcal{D}\Psi_3 - \mathcal{D}\Phi_{21} + \mathcal{P}\Phi_{22} + 2\mathcal{P}'\Lambda = \sigma\Psi_4 - 2\tau'\Psi_3 + 3\rho'\Psi_2 \tag{3.25b'}$$
$$- 3\kappa'\Psi_1 - \bar{\rho}\Phi_{22} - 2\bar{\tau}'\Phi_{21} - 2\tau'\Phi_{12} + 2\rho'\Phi_{11} + \bar{\sigma}'\Phi_{20}$$

$$\mathcal{P}\Psi_4 - \mathcal{D}'\Psi_3 - \mathcal{D}'\Phi_{21} + \mathcal{P}'\Phi_{20} = 3\sigma'\Psi_2 - 4\tau'\Psi_3 + \rho\Psi_4$$
$$- 2\kappa'\Phi_{10} + 2\sigma'\Phi_{11} + \bar{\rho}'\Phi_{20} - 2\bar{\tau}\Phi_{21} + \bar{\sigma}\Phi_{22} \tag{3.25c}$$

$$\mathcal{P}'\Psi_0 - \mathcal{D}\Psi_1 - \mathcal{D}\Phi_{01} + \mathcal{P}\Phi_{02} = 3\sigma\Psi_2 - 4\tau\Psi_1 + \rho'\Psi_0 \tag{3.25c'}$$
$$- 2\kappa\Phi_{12} + 2\sigma\Phi_{11} + \bar{\rho}\Phi_{02} - 2\bar{\tau}'\Phi_{01} + \bar{\sigma}'\Phi_{00}$$

$$\mathcal{P}\Psi_3 - \mathcal{D}'\Psi_2 - \mathcal{P}\Phi_{21} + \mathcal{D}\Phi_{20} - 2\mathcal{D}'\Lambda = 2\sigma'\Psi_1 - 3\tau'\Psi_2 + 2\rho\Psi_3$$
$$- \kappa\Psi_4 - 2\rho'\Phi_{10} + 2\tau'\Phi_{11} + \bar{\tau}'\Phi_{20} - 2\bar{\rho}\Phi_{21} + \bar{\kappa}\Phi_{22} \tag{3.25d}$$

$$\mathcal{P}'\Psi_1 - \mathcal{D}\Psi_2 - \mathcal{P}'\Phi_{01} + \mathcal{D}'\Phi_{02} - 2\mathcal{D}\Lambda = 2\sigma\Psi_3 - 3\tau\Psi_2 + 2\rho'\Psi_1 \tag{3.25d'}$$
$$- \kappa'\Psi_0 - 2\rho\Phi_{12} + 2\tau\Phi_{11} + \bar{\tau}\Phi_{02} - 2\bar{\rho}'\Phi_{01} + \bar{\kappa}'\Phi_{00}$$

GHP Contracted Bianchi Identities

$$\mathcal{P}\Phi_{11} - \mathcal{P}'\Phi_{00} - \mathcal{D}\Phi_{10} - \mathcal{D}'\Phi_{01} + 3\mathcal{P}\Lambda = (\rho' + \bar{\rho}')\Phi_{00} + 2(\rho + \bar{\rho})\Phi_{11}$$
$$- (\tau' + 2\bar{\tau})\Phi_{01} - (2\tau + \bar{\tau}')\Phi_{10} - \bar{\kappa}\Phi_{12} - \kappa\Phi_{21} + 2\sigma\Phi_{20} + \bar{\sigma}\Phi_{02} \tag{3.26a}$$

$$\mathcal{P}'\Phi_{11} - \mathcal{P}\Phi_{22} - \mathcal{D}'\Phi_{12} - \mathcal{D}\Phi_{21} + 3\mathcal{P}'\Lambda = (\rho + \bar{\rho})\Phi_{22} + 2(\rho' + \bar{\rho}')\Phi_{11} \quad (3.26a')$$
$$- (\tau + 2\bar{\tau}')\Phi_{21} - (2\tau' + \bar{\tau})\Phi_{12} - \bar{\kappa}'\Phi_{10} - \kappa'\Phi_{01} + 2\sigma'\Phi_{02} + \bar{\sigma}'\Phi_{20}$$

$$\mathcal{P}\Phi_{12} - \mathcal{P}'\Phi_{01} - \mathcal{D}\Phi_{11} - \mathcal{D}'\Phi_{02} + 3\mathcal{D}\Lambda = (\rho' + 2\bar{\rho}')\Phi_{01} + (2\rho + \bar{\rho})\Phi_{12} \quad (3.26b)$$
$$- (\tau' + \bar{\tau})\Phi_{02} - 2(\tau + \bar{\tau}')\Phi_{11} - \bar{\kappa}'\Phi_{00} - \kappa\Phi_{22} + \sigma\Phi_{21} + \bar{\sigma}'\Phi_{10}$$

$$\mathcal{P}'\Phi_{10} - \mathcal{P}\Phi_{21} - \mathcal{D}'\Phi_{11} - \mathcal{D}\Phi_{20} + 3\mathcal{D}'\Lambda = (\rho + 2\bar{\rho})\Phi_{21} + (2\rho' + \bar{\rho}')\Phi_{10} \quad (3.26b')$$
$$- (\tau + \bar{\tau}')\Phi_{20} - 2(\tau' + \bar{\tau})\Phi_{11} - \bar{\kappa}\Phi_{22} - \kappa'\Phi_{00} + \sigma'\Phi_{01} + \bar{\sigma}\Phi_{12}$$

The contents of the vacuum Einstein field equations can be obtained by putting Φ_{AB} and Λ equal to zero in Eqs. (3.23) and (3.24). The Bianchi identities (3.25) in this case have the following form.

GHP Vacuum Bianchi Identities

$$\mathcal{P}\Psi_1 - \mathcal{D}'\Psi_0 = -\tau'\Psi_0 + 4\rho\Psi_1 - 3\kappa\Psi_2 \quad (3.27a)$$

$$\mathcal{P}'\Psi_3 - \mathcal{D}\Psi_4 = -\tau\Psi_4 + 4\rho'\Psi_3 - 3\kappa'\Psi_2 \quad (3.27a')$$

$$\mathcal{P}\Psi_2 - \mathcal{D}'\Psi_1 = \sigma'\Psi_0 - 2\tau\Psi_1 + 3\rho\Psi_2 - 3\kappa\Psi_3 \quad (3.27b)$$

$$\mathcal{P}'\Psi_2 - \mathcal{D}\Psi_3 = \sigma\Psi_4 - 2\tau'\Psi_3 + 3\rho'\Psi_2 - 3\kappa'\Psi_1 \quad (3.27b')$$

$$\mathcal{P}\Psi_4 - \mathcal{D}'\Psi_3 = 3\sigma'\Psi_2 - 4\tau'\Psi_3 + \rho\Psi_4 \quad (3.27c)$$

$$\mathcal{P}'\Psi_0 - \mathcal{D}\Psi_1 = 3\sigma\Psi_2 - 4\tau\Psi_1 + \rho'\Psi_0 \quad (3.27c')$$

$$\mathcal{P}\Psi_3 - \mathcal{D}'\Psi_2 = 2\sigma'\Psi_1 - 3\tau'\Psi_2 + 2\rho\Psi_3 - \kappa\Psi_4 \quad (3.27d)$$

$$\mathcal{P}'\Psi_1 - \mathcal{D}\Psi_2 = 2\sigma\Psi_3 - 3\tau\Psi_2 + 2\rho'\Psi_1 - \kappa'\Psi_0 \quad (3.27d')$$

3.4 The Geometry of the Null Congruences

There are several ways to obtain the information about the structure of the gravitational field. One of them is to study the behaviour of different test particles in the field. If the massless particles (photons, neutrinos) are considered then the trajectories corresponding to them are light rays. It is seen that from the known trajectories of massless test particles (photons) the spacetime metric can be recovered uniquely up to a conformal transformation. It is also known that the study of light rays and congruence of light rays play an important role in the propagation of gravitational radiation. Moreover, when the problem of algebraic classification of the electromagnetic field and Weyl tensor of a gravitational field (or, other massless fields with spin) are considered, it turns out that with every suitable tensor there is associated a set of its eigen light-like vectors. Therefore, the corresponding tensor field (electromagnetic and gravitational) generate light-like vector fields in spacetime. The integral curves of a light-like vector field form a congruence of null curves [by the term *congruence* we mean a family of curves (or, surfaces, etc.) of which exactly one passes through each point of a spacetime region under consideration].

The study of the null congruences originated in the works of Ehlers and Sachs [2, 8, 9] and Newman and Penrose [6] where it was shown, along with some other important results, that there is a set of invariants (now known as optical scalars) which determine the geometric properties of the congruence of null curves and have a simple physical meaning.

Due to the all-important role of the congruence of null curves in general relativity, it is worthwhile to study the geometric properties of such congruences and the present section is devoted to this activity, using GHP formalism.

From the definition of the spin-coefficients [cf. Eq. (3.10)], we obtain

$$\nabla_a l_b = -(\mathcal{P}' - D')l_a l_b - \bar{\tau} l_a m_b - \tau l_a \bar{m}_b - (\mathcal{P} - D)n_a l_b - \bar{\kappa} n_a m_b - \kappa n_a \bar{m}_b$$
$$+ (\mathcal{D}' - \delta')m_a l_b + \bar{\sigma} m_a m_b + \rho m_a \bar{m}_b + (\mathcal{D} - \delta)\bar{m}_a l_b + \bar{\rho}\bar{m}_a m_b + \sigma \bar{m}_a \bar{m}_b$$

$$(3.28)$$

$$\nabla_a n_b = (\mathcal{P}' - D')l_a n_b - \kappa' l_a m_b - \bar{\kappa}' l_a \bar{m}_b - (\mathcal{P} - D)n_a n_b - \tau' n_a m_b - \bar{\tau}' n_a \bar{m}_b$$
$$- (\mathcal{D}' - \delta')m_a n_b + \sigma' m_a m_b + \bar{\rho}' m_a \bar{m}_b - (\mathcal{D} - \delta)\bar{m}_a n_b + \rho' \bar{m}_a m_b + \bar{\sigma}' \bar{m}_a \bar{m}_b$$

$$(3.29)$$

$$\nabla_a m_b = -\bar{\kappa}' l_a l_b - \tau l_a n_b + (\mathcal{P}' - D')l_a m_b - \bar{\tau}' n_a l_b - \kappa n_a n_b + (\mathcal{P} - D)n_a m_b$$
$$+ \bar{\rho}' m_a l_b + \rho m_a n_b + (\mathcal{D}' - \delta')m_a m_b + \bar{\sigma}' \bar{m}_a l_b + \sigma \bar{m}_a n_b + (\mathcal{D} - \delta)\bar{m}_a m_b$$

$$(3.30)$$

$$\nabla_a \bar{m}_b = -\kappa' l_a l_b - \bar{\tau} l_a n_b + (\mathcal{P}' - D')l_a \bar{m}_b - \tau' n_a l_b - \bar{\kappa} n_a n_b + (\mathcal{P} - D)n_a \bar{m}_b$$
$$+ \rho' \bar{m}_a l_b + \bar{\rho}\bar{m}_a n_b + (\mathcal{D} - \delta)\bar{m}_a \bar{m}_b + \sigma' m_a l_b + \bar{\sigma} m_a n_b + (\mathcal{D}' - \delta')m_a \bar{m}_b$$

$$(3.31)$$

where the derivatives D, D', δ and δ are defined as

$$D = l^a \nabla_a , \quad D' = n^a \nabla_a , \quad \delta = m^a \nabla_a , \quad \delta' = \bar{m}^a \nabla_a \qquad (3.32)$$

The contractions of Eqs. (3.28)–(3.31) yield

$$\nabla_a l^a = -(\rho + \bar{\rho}) - (\mathcal{P} - D) \qquad (3.33)$$

$$\nabla_a n^a = -(\rho' - \bar{\rho}') + (\mathcal{P}' - D') \qquad (3.34)$$

$$\nabla_a m^a = -(\bar{\tau}' + \tau) - (\mathcal{D} - \delta) \qquad (3.35)$$

$$\nabla_a \bar{m}^a = -(\tau' + \bar{\tau}) - (\mathcal{D}' - \delta') \qquad (3.36)$$

Equations (3.28)–(3.31) also lead to the following relations:

$$\nabla_{[a}l_{b]} = -2(\mathcal{P} - D)l_{[a}n_{b]} - (\bar{\tau} - \mathcal{D}' + \delta')l_{[a}m_{b]} - (\tau - \mathcal{D} + \delta)l_{[a}\bar{m}_{b]}$$
$$- \bar{\kappa}n_{[a}m_{b]} - \kappa n_{[a}\bar{m}_{b]} + 2i(\rho - \bar{\rho})m_{[a}\bar{m}_{b]} \tag{3.37}$$

$$\nabla_{[a}n_{b]} = 2(\mathcal{P} - D)l_{[a}n_{b]} - \kappa'l_{[a}m_{b]} - \bar{\kappa}'l_{[a}\bar{m}_{b]} - (-\tau - \mathcal{D}' + \delta')n_{[a}m_{b]}$$
$$+ (-\bar{\tau}' - \mathcal{D} + \delta)n_{[a}\bar{m}_{b]} - 2i(\rho' - \bar{\rho}')m_{[a}\bar{m}_{b]} \tag{3.38}$$

$$\nabla_{[a}m_{b]} = -(\tau - \bar{\tau}')l_{[a}n_{b]} - \{2i(\mathcal{P}' - D') + \bar{\rho}'\}l_{[a}m_{b]} - \bar{\sigma}'l_{[a}\bar{m}_{b]}$$
$$- \{2i(\mathcal{P} - D) + \rho\}n_{[a}m_{b]} - \sigma n_{[a}\bar{m}_{b]} - (\mathcal{D} - \delta)m_{[a}\bar{m}_{b]} \tag{3.39}$$

$$\nabla_{[a}\bar{m}_{b]} = -(\bar{\tau} - \tau')l_{[a}n_{b]} - \{2i(\mathcal{P}' - D') + \rho'\}l_{[a}\bar{m}_{b]} - \sigma'l_{[a}m_{b]}$$
$$- \{2i(\mathcal{P} - D) + \bar{\rho}\}n_{[a}\bar{m}_{b]} - \bar{\sigma}n_{[a}m_{b]} - (\mathcal{D}' - \delta')\bar{m}_{[a}m_{b]} \tag{3.40}$$

When the GHP derivative operators given by Eq. (3.19) act on the tetrad vectors $l^a, n^a, m^a, \bar{m}^a$, they give rise to the following set of equations:

$$(a)\ \mathcal{P}l^a = -\kappa\bar{m}^a - \bar{\kappa}m^a\ ,\ (b)\ \mathcal{P}'l^a = -\bar{\tau}m^a - \tau\bar{m}^a$$

$$(c)\ \mathcal{D}l^a = -\bar{\rho}m^a - \sigma\bar{m}^a\ ,\ (d)\ \mathcal{D}'l^a = -\bar{\sigma}m^a - \rho\bar{m}^a \tag{3.41}$$

$$(a)\ \mathcal{P}n^a = -\tau'm^a - \bar{\tau}'\bar{m}^a\ ,\ (b)\ \mathcal{P}'n^a = -\kappa'm^a - \bar{\kappa}'\bar{m}^a$$

$$(c)\ \mathcal{D}n^a = -\rho'm^a - \bar{\sigma}'\bar{m}^a\ ,\ (d)\ \mathcal{D}'n^a = -\bar{\rho}'\bar{m}^a - \sigma'm^a \tag{3.42}$$

$$(a)\ \mathcal{P}m^a = -\bar{\tau}'l^a - \kappa n^a\ ,\ (b)\ \mathcal{P}'m^a = -\bar{\kappa}'l^a - \tau n^a$$

$$(c)\ \mathcal{D}m^a = -\bar{\sigma}'l^a - \sigma n^a\ ,\ (d)\ \mathcal{D}'m^a = -\bar{\rho}'l^a - \rho n^a \tag{3.43}$$

$$(a)\ \mathcal{P}\bar{m}^a = -\tau'l^a - \bar{\kappa}n^a\ ,\ (b)\ \mathcal{P}'\bar{m}^a = -\kappa'l^a - \bar{\tau}n^a$$

$$(c)\ \mathcal{D}\bar{m}^a = -\sigma'l^a - \bar{\sigma}n^a\ ,\ (d)\ \mathcal{D}'\bar{m}^a = -\rho'l^a - \bar{\rho}n^a \tag{3.44}$$

In GHP formalism, the role of the covariant derivative operator is replaced by the operator Θ_a defined by Eq. (3.19b). Applying this operator on tetrad vectors, we get

$$\Theta_a l_b = (-\bar{\tau}l_a - \bar{\kappa}n_a + \bar{\sigma}m_a + \bar{\rho}\bar{m}_a)m_b + (-\tau l_a - \kappa n_a + \rho m_a + \sigma\bar{m}_a)\bar{m}_b \tag{3.45}$$

$$\Theta_a n_b = (-\kappa'l_a - \tau'n_a + \sigma'm_a + \rho'\bar{m}_a)m_b + (-\bar{\kappa}'l_a - \bar{\tau}'n_a + \bar{\rho}'m_a + \bar{\sigma}'\bar{m}_a)\bar{m}_b \tag{3.46}$$

$$\Theta_a m_b = (-\bar{\kappa}'l_a - \bar{\tau}'n_a + \bar{\rho}'m_a + \bar{\sigma}'\bar{m}_a)l_b + (-\tau l_a - \kappa n_a + \rho m_a + \sigma\bar{m}_a)n_b \tag{3.47}$$

$$\Theta_a\bar{m}_b = (-\kappa'l_a - \tau'n_a + \rho'\bar{m}_a + \sigma'm_a)l_b + (-\bar{\tau}l_a - \bar{\kappa}n_a + \bar{\rho}\bar{m}_a + \bar{\sigma}m_a)n_b \tag{3.48}$$

so that

$$\Theta_a l^a \;=\; -(\rho + \bar\rho) \tag{3.49}$$

$$\Theta_a n^a \;=\; -(\rho' + \bar\rho') \tag{3.50}$$

$$\Theta_a m^a \;=\; -(\tau + \bar\tau') \tag{3.51}$$

$$\Theta_a \bar m^a \;=\; -(\bar\tau + \tau') \tag{3.52}$$

From Eqs. (3.41) and (3.19b), we also have

$$l_a\Theta_b l_c = l_a l_b(-\bar\tau m_c - \tau\bar m_c) + l_a n_b(-\kappa\bar m_c - \bar\kappa m_c)$$
$$- l_a m_b(-\bar\sigma m_c - \rho\bar m_c) - l_a \bar m_b(-\bar\rho m_c - \sigma\bar m_c) \tag{3.53}$$

$$l_b\Theta_a l_c = l_b l_a(-\bar\tau m_c - \tau\bar m_c) + l_b n_a(-\kappa\bar m_c - \bar\kappa m_c)$$
$$- l_b m_a(-\bar\sigma m_c - \rho\bar m_c) - l_b \bar m_a(-\bar\rho m_c - \sigma\bar m_c) \tag{3.54}$$

$$l_b\Theta_c l_a = l_b l_c(-\bar\tau m_a - \tau\bar m_a) + l_b n_c(-\kappa\bar m_a - \bar\kappa m_a)$$
$$- l_b m_c(-\bar\sigma m_a - \rho\bar m_a) - l_b \bar m_c(-\bar\rho m_a - \sigma\bar m_a) \tag{3.55}$$

$$l_c\Theta_a l_b = l_c l_a(-\bar\tau m_b - \tau\bar m_b) + l_c n_a(-\kappa\bar m_b - \bar\kappa m_b)$$
$$- l_c m_a(-\bar\sigma m_b - \rho\bar m_b) - l_c \bar m_a(-\bar\rho m_b - \sigma\bar m_b) \tag{3.56}$$

$$l_c\Theta_b l_a = l_c l_b(-\bar\tau m_a - \tau\bar m_a) + l_c n_b(-\kappa\bar m_a - \bar\kappa m_a)$$
$$- l_c m_b(-\bar\sigma m_a - \rho\bar m_a) - l_c \bar m_b(-\bar\rho m_a - \sigma\bar m_a) \tag{3.57}$$

$$l_a\Theta_c l_b = l_a l_c(-\bar\tau m_b - \tau\bar m_b) + l_a n_c(-\kappa\bar m_b - \bar\kappa m_b)$$
$$- l_a m_c(-\bar\sigma m_b - \rho\bar m_b) - l_a \bar m_c(-\bar\rho m_b - \sigma\bar m_b) \tag{3.58}$$

so that, after simplifications, we have

$$l_{[a}\Theta_b l_{c]} = \frac{1}{3!}\{-\kappa l_a n_b\bar m_c - \bar\kappa l_a n_b m_c + \rho l_a m_b\bar m_c + \bar\rho l_a \bar m_b m_c$$
$$+ \kappa l_a n_a\bar m_c + \bar\kappa l_b n_a m_c - \rho l_b m_a\bar m_c - \bar\rho l_b \bar m_a m_c$$
$$- \kappa l_b n_c\bar m_a - \bar\kappa l_b n_c m_a + \rho l_b m_c\bar m_a + \bar\rho l_b \bar m_c m_a$$
$$+ \kappa l_c n_b\bar m_a + \bar\kappa l_c n_b m_a - \rho l_c m_b\bar m_a - \bar\rho l_c \bar m_b m_a$$
$$- \kappa l_c n_a\bar m_b - \bar\kappa l_c n_a m_b + \rho l_c m_a\bar m_b + \bar\rho l_c \bar m_a m_b$$
$$+ \kappa l_a n_c\bar m_b + \bar\kappa l_a n_c m_b - \rho l_a m_c\bar m_b - \bar\rho l_a \bar m_c m_b\} \tag{3.59}$$

We also have

$$l^a\Theta_a l^b = \mathcal{P}l^b \;,\;\; l^a\Theta_a n^b = \mathcal{P}n^b \;,\;\; l^a\Theta_a m^b = \mathcal{P}m^b \;,\;\; l^a\Theta_a \bar m^b = \mathcal{P}\bar m^b \tag{3.60}$$

From Eqs. (3.28) and (3.41), it can easily be shown that

$$\nabla_a = l_a D' + n_a D - m_a \delta' - \bar{m}_a \delta \tag{3.61}$$

where the derivatives D, D', δ and δ' are defined by Eq. (3.32).

The duals of the skew products are

$$(l_{[a}n_{b]})^* = i \, m_{[a}\bar{m}_{b]} \,, \; (l_{[a}m_{b]})^* = - i \, l_{[a}m_{b]}$$

$$(l_{[a}\bar{m}_{b]})^* = i \, l_{[a}\bar{m}_{b]} \,, \; (n_{[a}m_{b]})^* = i \, n_{[a}m_{b]} \tag{3.62}$$

$$(n_{[a}\bar{m}_{b]})^* = - i \, n_{[a}\bar{m}_{b]} \,, \; (m_{[a}\bar{m}_{b]})^* = i \, l_{[a}n_{b]}$$

Using Eq. (3.62), the dual of Eq. (3.37) is given by

$$(\nabla_{[a}l_{b]})^* = - 2i(\mathcal{P} - D)m_{[a}\bar{m}_{b]} + i(\bar{\tau} - \mathcal{D}' + \delta')l_{[a}m_{b]} - i(\tau - \mathcal{D} + \delta)l_{[a}\bar{m}_{b]}$$
$$- i\bar{\kappa}n_{[a}m_{b]} + i\kappa n_{[a}\bar{m}_{b]} - 2(\rho - \bar{\rho})l_{[a}n_{b]} \tag{3.63}$$

Thus, from Eqs. (3.37) and (3.63), we have

$$\nabla_{[a}l_{b]} - i(\nabla_{[a}l_{b]})^* = \{-2(\mathcal{P} - D) + 2i(\rho - \bar{\rho})\}(l_{[a}n_{b]} + m_{[a}\bar{m}_{b]})$$
$$- 2(\tau - \mathcal{D} + \delta)l_{[a}\bar{m}_{b]} - 2\bar{\kappa}n_{[a}m_{b]} - 2\kappa n_{[a}\bar{m}_{b]} \tag{3.64}$$

Now using the properties of the tetrad $Z^a_\mu = \{l^a, n^a, m^a, \bar{m}^a\}$ [cf. Eqs. (3.1)–(3.3)] in Eqs. (3.49) and (3.64), we get after simplification

$$\{\nabla_{[a}l_{b]} - i(\nabla_{[a}l_{b]})^*\}l^b = -(\nabla_c l^c + 2\rho)l_a + 2\mathcal{P}l_a \tag{3.65}$$

where $\mathcal{P}l_a$ is given by Eq. (3.41a).

We shall now prove a number of theorems illustrating the geometric properties of the spin-coefficients (scalars) appearing in the GHP formalism (see also [1]).

Theorem 3.1 *The null congruence $C(l)$ is geodesic if and only if $\kappa = 0$ and by a suitable scaling we may set $\mathcal{P} - D = 0$.*

Theorem 3.2 *The null congruence $C(n)$ is geodesic if and only if $\kappa' = 0$ and by a suitable scaling we may set $\mathcal{P}' - D' = 0$.*

The proofs of Theorems 3.1 and 3.2 follow from Eqs. (3.41a) and (3.42a).

Theorem 3.3 *If we choose $\mathcal{P} - D = 0$, then the tetrad $\{l^a, n^a, m^a, \bar{m}^a\}$ is parallelly propagated along $C(l)$ when $\kappa = \tau' = 0$.*

Proof The parallel displacement along $C(l)$ requires that $l^a \Theta_a Z^b_\mu = 0$ for each μ, i.e. $\mathcal{P}l^b = \mathcal{P}n^b = \mathcal{P}m^b = \mathcal{P}\bar{m}^b = 0$. From Theorem 3.1, $\mathcal{P}l^a = 0$ implies that $\kappa = 0$ and thus from Eqs. (3.43a) and (3.44a) we have $\tau' = 0$ which proves the theorem.

Similarly, we have the following theorems.

Theorem 3.4 *If we choose $\mathcal{P}' - D' = 0$, then the tetrad $\{l^a, n^a, m^a, \bar{m}^a\}$ is parallelly propagated along $C(n)$ when $\kappa' = \tau = 0$.*

Theorem 3.5 *The null congruence $C(l)$ is geodesic if and only if*

$$l^{[b}\mathcal{P}l^{a]} = 0$$

Proof From Eq. (3.41a) we have

$$l^b\mathcal{P}l^a = -l^b\kappa\bar{m}^a - l^b\bar{\kappa}m^a \text{ and } l^a\mathcal{P}l^b = -l^a\kappa\bar{m}^b - l^a\bar{\kappa}m^b$$

which leads to

$$l^{[b}\mathcal{P}l^{a]} = -\kappa l^{[b}\bar{m}^{a]} - \bar{\kappa}l^{[b}m^{a]}$$

Since the linear dependence of the bi-vectors $l^{[b}m^{a]}$ and $l^{[b}\bar{m}^{a]}$ imply the vanishing of κ and $\bar{\kappa}$, therefore, $l^{[b}\mathcal{P}l^{a]} = 0$.

In a similar manner for $C(n)$, we have

Theorem 3.6 *The null congruence $C(n)$ is geodesic if and only if*

$$n^{[b}\mathcal{P}'n^{a]} = 0$$

We shall now prove the following lemma for our later investigation.

Lemma 3.1 *Let $C(l)$ be a null geodesic congruence then $l_{[a}\Theta_b l_{c]} = 0$ is equivalent to $\rho = \bar{\rho}$.*

Proof If $C(l)$ is a null geodesic congruence then Eq. (3.59) leads to

$$l_{[a}\Theta_b l_{c]} = (\rho - \bar{\rho})l_{[a}m_b\bar{m}_{c]} \tag{3.66}$$

As the vectors l_a, n_a, m_a and $\bar{m}_a$ are linearly independent, the right-hand side of Eq. (3.66) is non-zero and thus $l_{[a}\Theta_b l_{c]} = 0$ implies that $\rho = \bar{\rho}$. Conversely, if $\rho = \bar{\rho}$ then from Eq. (3.66), $l_{[a}\Theta_b l_{c]} = 0$ which completes the proof of the lemma.

From Schouten [10], the condition of Lemma 3.1 (in the language of GHP formalism), viz. $k_{[a}\Theta_b k_{c]} = 0$ is the necessary and sufficient condition for the congruence $C(k)$ to be hypersurface forming (or hypersurface orthogonal)[1] and thus from Lemma 3.1, we have the following theorems.

Theorem 3.7 *The null vector field l is hypersurface orthogonal if and only if $\kappa = 0$ and $\rho = \bar{\rho}$.*

Theorem 3.8 *The null vector field n is hypersurface orthogonal if and only if $\kappa' = 0$ and $\rho' = \bar{\rho}'$.*

From Eqs. (3.41a) and (3.65), we also have the following theorem.

---

[1]In fact, the necessary and sufficient condition for the congruence $C(k)$ to be hypersurface orthogonal, according to Schouten [10], is $k_{[a}\nabla_b k_{c]} = 0$.

Theorem 3.9 *The null congruence $C(l)$ is geodesic if and only if*

$$\{\nabla_{[a}l_{b]} - i(\nabla_{[a}l_{b]})^*\}l^b = -(\nabla_c l^c + 2\rho)l_a$$

Moreover, from the properties of the tetrad $Z^a_\mu = \{l^a, n^a, m^a, \bar{m}^a\}$ and Eq. (3.63) it can easily be verified that

$$(\nabla_{[a}l_{b]})^* l^b = -\Im(\rho)l_a - \frac{i}{2}\mathcal{P}l_a \tag{3.67a}$$

$$(\nabla_{[a}l_{b]})^* n^b = \Im(\rho)n_a + \frac{i}{2}\mathcal{P}'l_a - \frac{i}{2}(\mathcal{D}' - \delta')m_a - \frac{i}{2}(\mathcal{D} - \delta)\bar{m}_a \tag{3.67b}$$

$$(\nabla_{[a}l_{b]})^* m^b = \frac{i}{2}(\mathcal{D} - \delta)l_a + \frac{i}{2}(\tau l_a - \kappa n_a) + i(\mathcal{P} - \mathcal{D})m_a \tag{3.67c}$$

$$(\nabla_{[a}l_{b]})^* \bar{m}^b = -\frac{i}{2}(\mathcal{D}' - \delta')l_a - \frac{i}{2}(\bar{\tau}l_a - \bar{\kappa}n_a) - i(\mathcal{P} - \mathcal{D})\bar{m}_a \tag{3.67d}$$

Further, if $C(l)$ is null geodesic congruence with $\mathcal{P} - \mathcal{D} = 0$, then l_a will be tangent vector corresponding to an affine parametrization and the scalars

$$\theta_{C(l)} = \frac{1}{2}\nabla_a l^a = -\Re(\rho) = -(\rho + \bar{\rho})$$

$$\omega_{C(l)} = -\{\frac{1}{2}\nabla_{[a}l_{b]}\nabla^a l^b\}^{1/2} = -\Im(\rho) = -(\rho - \bar{\rho})$$

and

$$\sigma\bar{\sigma} = |\sigma| = \frac{1}{2}\{\nabla_{[a}l_{b]}\nabla^a l^b - (\theta_{C(l)})^2\}^{1/2}$$

are, respectively, the *expansion*, the *twist* and the *shear* of the congruence $C(l)$.

From Eqs. (3.67), we thus have the following theorem.

Theorem 3.10 *If $C(l)$ is a null geodesic congruence with $\mathcal{P} - \mathcal{D} = 0$ then*

$$(\nabla_{[a}l_{b]})^* l^b = \omega_{C(l)}l_a = -\Im(\rho)l_a$$

$$(\nabla_{[a}l_{b]})^* n^b = \omega_{C(l)}n_a + \frac{i}{2}\{\mathcal{P}'l_a - (\mathcal{D}' - \delta')m_a + (\mathcal{D} - \delta)\bar{m}_a\}$$

$$(\nabla_{[a}l_{b]})^* m^b = \frac{i}{2}(\mathcal{D} - \delta + \tau)l_a$$

$$(\nabla_{[a}l_{b]})^* \bar{m}^b = -\frac{i}{2}(\mathcal{D}' - \delta' - \bar{\tau})l_a$$

We conclude our investigations about the null congruences by saying that the null geodesic congruences $C(l)$ and $C(n)$ are shear-free if $\sigma = 0$ and $\sigma' = 0$, respectively.

References

1. Ahsan, N.: A study of compacted spin-coefficient formalism in general relativity. Ph.D Thesis, Aligarh Muslim University, Aligarh, India (2000)
2. Ehlers, J., Sachs, R.K.: Z. Physik **155**, 498 (1959)
3. Geroch, R., Held, A., Penrose, R.: J. Math. Phys. **14**, 874–881 (1973)
4. Held, A.: J. Math. Phys. **17**, 39–45 (1976)
5. Held, A.: Gen. Rel. Grav. **7**, 177–198 (1976)
6. Newman, E.T., Penrose, R.: J. Math. Phys. **3**, 566–578 (1962)
7. Penrose, R., Rindler, W.: Spinors and Spacetime. Cambridge University Press, New York (1984)
8. Sachs, R.K.: Proc. Roy. Soc. Lond. A **264**, 309 (1961)
9. Sachs, R.K.: Proc. Roy. Soc. Lond. A **270** (1962)
10. Schouten, J.A.: Ricci Calculus, 2nd edn. Springer (1954)

Chapter 4
Lanczos Potential and Tetrad Formalism

4.1 Introduction

Consider a four-dimensional spacetime endowed with a metric g_{ij}. For the vector field v_k, the Riemann curvature tensor $R^h{}_{ijk}$ is defined through the Ricci identity

$$v_{k;i;j} - v_{k;j;i} = R^h{}_{ijk} v_h \tag{4.1}$$

The Riemann curvature tensor can be decomposed as follows [2]:

$$R_{hijk} = C_{hijk} + E_{hijk} + G_{hijk} \tag{4.2}$$

where

$$E_{hijk} = \frac{1}{2}(g_{hj}S_{ik} + g_{ik}S_{hj} - g_{hk}S_{ij} - g_{ij}S_{hk}) \tag{4.3}$$

$$G_{hijk} = \frac{R}{12}(g_{hj}g_{ik} - g_{hk}g_{ij}) \tag{4.4}$$

$$S_{ij} = R_{ij} - \frac{1}{4}Rg_{ij} \tag{4.5}$$

$$R_{ij} = R_{ikj}{}^k , \quad R = R_i^i \tag{4.6}$$

The irreducible tensor[1] C_{hijk} is the Weyl tensor and satisfies the same algebraic properties as that of the Riemann tensor

$$C_{hijk} = -C_{ihjk} = -C_{hikj} = C_{jkhi} , \quad C_{h[ijk]} = 0 \tag{4.7}$$

---

[1] A tensor quantity is said to be reducible if it can be decomposed into parts which transform along themselves. If such a decomposition is not possible, then the quantity is called irreducible (cf., [2]).

© Springer Nature Singapore Pte Ltd. 2019
Z. Ahsan, *The Potential of Fields in Einstein's Theory of Gravitation*,
https://doi.org/10.1007/978-981-13-8976-4_4

The other parts in the decomposition (4.2) have the same symmetries. Moreover,

$$C^t{}_{itj} = 0 \, , \; E^t{}_{itj} = S_{ij} \, , \; G^t{}_{itj} = \frac{1}{4} g_{ij} \, R \qquad (4.8)$$

The Weyl tensor has ten independent components and is completely traceless, i.e. the contraction with respect to each pair of indices vanishes. A spacetime is said to be conformally flat if $C_{hijk} = 0$.

A consequence of the Eqs. (4.7) and (4.8) is the mixed dual property

$$^*C_{hijk} = C^*_{hijk} \qquad (4.9)$$

where

$$^*C_{hijk} = \frac{1}{2} \eta_{hirs} C^{rs}{}_{jk} \, , \; C^*_{hijk} = \frac{1}{2} \eta_{jkrs} C_{hi}{}^{rs} \qquad (4.10)$$

are, respectively, the left and right duals of the Weyl tensor.

When Einstein's vacuum field equations

$$R_{ij} = 0 \qquad (4.11)$$

are imposed then from Eq. (4.2) all that remains of the gravitational field is the Weyl tensor and it (Weyl tensor) describes the pure gravitational field.

4.2   Lanczos Potential

Lanczos had a deep interest in general theory of relativity and its study by means of variational principles. He explored many modifications of the Einstein field equations based on Lagrangian functions which were quadratic in the components of the Riemann curvature tensor and whose Euler equations were therefore different from the Einstein ones since the latter may be obtained by using the scalar curvature alone. In his 1962 paper [32] entitled 'The Splitting of the Riemann Tensor', he returned to the study of the algebraically independent parts of the Riemann tensor and to the study of a quadratic Lagrangian whose variation vanishes identically. He paid particular attention to a part of this tensor which he called self-dual and which satisfied other conditions. This part is the conformal or Weyl tensor but Lanczos did not use this fact. He was mainly concerned with finding a 'generating function' for the conformal tensor. He used the discussion of the variational principle mentioned above to express the conformal tensor in terms of the derivatives of a tensor L_{ijk}. He also discussed the equations which this tensor has to satisfy in case the conformal tensor vanished.

Because the tensors E_{hijk} and G_{hijk} are derived from simpler irreducible tensors with fewer indices, namely S_{ij} and R, Lanczos [32] thought that the Weyl tensor can also be derivable from a simpler tensor field, and this indeed can be done through the covariant differentiation of a tensor field L_{ijk}. This tensor field satisfies the following symmetries

$$(40 \text{ conditions}) \quad L_{ijk} = -L_{jik} \tag{4.12}$$

$$(4 \text{ conditions}) \quad L_i{}^t{}_t = 0 \quad (\text{or, } g^{kl}L_{kil} = 0) \tag{4.13}$$

$$(4 \text{ conditions}) \quad L_{ijk} + L_{jki} + L_{kij} = 0 \; (\text{or, } {}^*L_i{}^t{}_t = 0) \tag{4.14}$$

Thus, the tensor field L_{ijk}, which has at most 64 independent components (in four dimensions), has been reduced to at most 16 independent components. In order to have a perfect match with the Weyl tensor, Lanczos imposed the six equations

$$L_{ij}{}^k{}_{;k} = 0 \tag{4.15}$$

and now the tensor field L_{ijk} has only ten effective degrees of freedom. Equation (4.15) is known as *Lanczos differential gauge condition* and is equivalent to

$$(L_{ij}^k + L_{ij}{}^{jk})_{;k} = 0 \tag{4.15a}$$

Using this tensor, Lanczos generated the gravitational field through the equation (c.f., [21, 32, 45])

$$C_{hijk} = L_{[hi][j;k]} + L_{[jk][h;i]} - {}^*L^*{}_{[hi][j;k]} - {}^*L^*{}_{[jk][h;i]} \tag{4.16}$$

where dual operation is applied to each pair of skew-symmetric brackets as indicated and the double dual is defined as ${}^*A^*{}_{hijk} = \frac{1}{4}\eta_{hilm}\eta_{jkno}A^{lmno}$.

Equation (4.16) can also be expressed as [21]

$$C_{hijk} = L_{hij;k} - L_{hik;j} + L_{jkh;i} - L_{jki;h} + L_{(hk)}g_{ij} + L_{(ij)}g_{hk} - L_{(hj)}g_{ik} - L_{(ik)}g_{hj}$$

$$+ \frac{2}{3}L^{pq}{}_{p;q}(g_{hj}g_{ik} - g_{hk}g_{ij}) \tag{4.17}$$

where

$$L_{hk} = L^t{}_{hk;t} - L^t{}_{ht;k} \tag{4.18}$$

and round bracket denotes symmetrization.

Using Eqs. (4.13), (4.15) and (4.18), (4.17) can also be written as

$$C_{hijk} = L_{hij;k} - L_{hik;j} + L_{jkh;i} - L_{jki;h} + \frac{1}{2}(L^p{}_{hk;p} + L^p{}_{kh;p})g_{ij} + \frac{1}{2}(L^p{}_{ij;p} + L^p{}_{ji;p})g_{hk}$$

$$- \frac{1}{2}(L^p{}_{hj;p} + L^p{}_{jh;p})g_{ik} - \frac{1}{2}(L^p{}_{ik;p} + L^p{}_{ki;p})g_{hj} \tag{4.19}$$

Equations (4.16)/(4.17)/(4.19) is known as *Weyl–Lanczos equation* and the tensor field L_{ijk} is now commonly known as *Lanczos potential* or *Lanczos spin tensor*. In this text, we shall consider the Weyl–Lanczos equation (4.19).

If F_{ij} denotes the electromagnetic field tensor then it is known that the electromagnetic field can be generated by a vector field A_i through the equation

$$F_{ij} = A_{i;j} - A_{j;i} \tag{4.20}$$

where A_i acts as the potential for the electromagnetic field. Moreover, F_{ij} satisfies Maxwell's equations

$$F_{ij;k} + F_{jk;i} + F_{ki;j} = 0 \tag{4.21}$$

$$F^{ij}{}_{;j} = J^i \tag{4.22}$$

Since the electromagnetic field is generated through the covariant differentiation of the vector field A_i, the gravitational field (the Weyl tensor C_{hijk}) can also be generated by the covariant differentiation of the tensor field L_{ijk}; and this what exactly has been given by Eq. (4.19)—the Weyl–Lanczos equation. The role of Lanczos potential L_{ijk} with respect to Weyl tensor C_{hijk} (gravitational field) is same as that of the vector potential A_i for the electromagnetic field tensor F_{ij}.

It is known that 20 Bianchi identities

$$R_{hijk;l} + R_{hikl;j} + R_{hilj;k} = 0 \tag{4.23}$$

are reducible to four equations

$$G^{ik}{}_{;k} = 0 \tag{4.24}$$

where

$$G_{ij} = R_{ij} - \frac{1}{2} R \, g_{ij} \tag{4.25}$$

is the Einstein tensor. Moreover, we have 16 irreducible equations

$$C_{ijk}{}^{l}{}_{;l} = J_{ijk} \tag{4.26}$$

where

$$J_{ijk} = R_{k[i;j]} - \frac{1}{6} g_{k[i} R_{;j]} \tag{4.27}$$

is known as the Schouten tensor. Equation (4.26) may be taken as the 'Maxwell-like' Eq. (4.22).

These equations suggest that like potential field L_{ijk}, the tensor field J_{ijk} is irreducible with at most 16 independent components because of its symmetries and only ten effective degrees of freedom because of the six identities

$$J_{ij}{}^k_{;k} = 0 \tag{4.28}$$

It may be noted that the right-hand side of Eq. (4.27) can be expressed entirely in terms of covariant derivatives of the irreducible tensor field S_{ij}, i.e. the traceless part of Ricci and Einstein tensors.

Using Eqs. (4.16) and (4.26), Dolan and Kim [21] obtained the following wave equation:

$$\Box L_{ijk} + R^t{}_{ipk}L_j{}^P{}_t - R^t{}_{jkp}L_i{}^P{}_t - R^t{}_{ijp}L_t{}^P{}_k + R_{tk}L_{ij}{}^t$$

$$- R_{tj}L_k{}^t{}_i + R_{ti}L_k{}^t{}_j + R^q{}_{jtp}L_q{}^{tp}g_{ik} - R^q{}_{itp}L_q{}^{tp}g_{jk} = J_{ijk} \tag{4.29}$$

where $\Box L_{ijk} \doteq g^{lm} L_{ijk;l;m}$. Equation (4.29) can also be expressed as [21]

$$\Box L_{ijk} + 2R^t{}_k L'_{ijt} - R_i{}^t L_{jkt} + R_j{}^t L_{ikt} - g_{ik}R^{pt}L_{pjt}$$

$$+ g_{jk}R^{pt}L_{pit} - \frac{1}{2}RL_{ijk} = J_{ijk} \tag{4.30}$$

which, in empty space, reduces to

$$\Box L_{ijk} = 0 \tag{4.31}$$

It may be noted that the 'Maxwell-like' Eq. (4.26) and Weyl–Lanczos relations (4.16) are sufficient to provide a close analogy between electromagnetic radiation via the 4-potential (vector A_i) and gravitational radiation via the Lanczos potential (tensor L_{ijk}). A comparison between the analogies of the electromagnetic and gravitation theories is given in Table 4.1 [21].

In 1962, Lanczos proved the existence of a tensor L_{ijk} as a potential to the Weyl tensor C_{ijkl}, but there was little development in the subject for quite some time. In 1975, Zund [47] has studied this potential using the spinor calculus, while in 1983 Bampi and Caviglia [17] proved the existence of Lanczos potential to a larger class of four tensors. Using spinor formalism, in 1988, Illge [31] proved the existence of Lanczos potential in four dimensions and obtained the wave equation for the Lanczos potential both in spinor and tensor forms. These wave equations were further studied, in 1994, by Dolan and Kim [21, 22] who gave a correct tensor version of the wave equation appeared in the paper of Illge [31]. A number of identities satisfied by the Lanczos spin tensor were obtained by Edgar [25] and Edgar and Höglund [26]. Novello and Velloso [42] obtained an algorithm for calculating the Lanczos potential for perfect fluid spacetimes. Dolan and Muratori [23] not only generalized the results of Novello and Velloso [42] but also made corrections in the earlier versions of the Weyl–Lanczos equations appeared in the literature (cf., [15, 47]). The Lanczos potential in Kerr geometry has been studied by Bergqvist [18], and a relationship between the Lanczos potential for vacuum spacetimes and the Ernst potential has been established by Dolan and Muratori [23]. For a larger class of spacetimes, using

Table 4.1 A comparison between electromagnetic and gravitational theories

Fields	F_{ij}	C_{hijk}
Potentials	A_i	L_{ijk}
Field relations	$F_{ij} = A_{i;j} - A_{j;i}$	$C_{hijk} = W(L)_{hijk}$
Gauge invariance	$A'_i = A_i + \chi_{,i}$ $W_{ij}(A') = W_{ij}(A)$	$L'_{ijk} = L_{ijk} + \chi_{ijk}$ $W_{hijk}(L') = W_{hijk}(L)$
Gauge conditions	$A^i{}_{;i} = 0$	$L_{ij}{}^t{}_{;t} = 0$
Field equations	$F^{ij}{}_{;j} = J^i$	$C_{ijk}{}^t{}_{;t} = J_{ijk}$
Potential wave equation in *matter*	$\Box A_i + R_i{}^k A_k = J_i$	$\Box L_{ijk} + 2R_k{}^t L_{ijt} - R_i{}^t L_{jkt}$ $-R_j{}^t L_{kit} - g_{ik}R^{pt}L_{pjt}$ $+g_{jk}R^{pt}L_{pit} - \frac{1}{2}RL_{ijk} = J_{ijk}$
Potential wave equation in *vacuo*	$\Box A_i + R_i{}^k A_k = 0$	$\Box L_{ijk} = 0$
Field wave equation in *matter*	$\Box F_{ij} + R^t{}_i F_{tj} - R^t{}_j F_{ti}$ $-2R_{risj}F^{rs} - F_i{}^t{}_{;t;j}$ $+F_j{}^t{}_{;t;i} = 0$	$\Box R_{hijk} + 4R_{hpq[j}R_{k]}{}^q{}_i{}^p$ $-R_{hipq}R^{pq}{}_{jk} + 2R^q{}_{[k}R_{j]qhi}$ $+2R_{j[i;h];k} + 2R_{k[h;i];j} = 0$
Field wave equation in *vacuo*	$\Box F_{ij} + R^t{}_i F_{tj} - R^t{}_j F_{ti} - 2R_{risj}F^{rs} = 0$	$\Box C_{hijk} + 4C_{hpq[j}C_{k]}{}^q{}_i{}^p - C_{hi}{}^{pq}C_{pqjk} = 0$

the spinor formalism, the Lanczos potential was obtained by Torris de Castillo [46], while using the Newman–Penrose formalism, Ares de Parga, Lopez-Bonilla, Gaftoi, and co-workers, in a series of papers [14, 27, 28, 30, 33] obtained the Lanczos potential for various algebraically special spacetimes. Massa and Pagani [39] have shown that in four dimensions, the Riemann tensor cannot, in general, be expressed in terms of a Lanczos potential. This result was later on generalized by Edgar [24] for n dimensions. While Andersson and Edgar [12] showed that the Lanczos potential can be defined in a very simple way directly from the spinor dyad and have obtained a link between the Lanczos potential and the spin-coefficients for some spacetime. In connection with the linear spin-2 theory and a Born–Infeld type action, the Lanczos potential has been studied, in 2003, by Cartin [20]. In 2007, Mora and Sanchez [41] used the wave equations given by Dolan and Kim [21, 22] to study the Lanczos potential. While a prescription for constructing a Lanczos potential for a cosmological model, which is a purely gravitational perturbation of FLRW spacetime, was given by Mena and Tod [40], which in turn led to a definition of gravitational entropy. Moreover, the idea of Lanczos spin tensor is very useful in the analysis of the Lienard-Wiechert field [1, 13, 16, 19, 28, 34–37]. Using NP and GHP formalisms, Ahsan et al. [3–8, 11] have given a general prescription to obtain the Lanczos potential for algebraically special spacetimes as well as perfect fluid spacetimes. While using the methods of local and isometric embedding, Lovelock's theorem and wave equations, the Lanczos potential for Gödel spacetime have been obtained by Ahsan et al. [9, 10]. Although the physical meaning of Lanczos tensor is not yet very clear (cf., [18, 27, 29, 32, 38, 43, 44], but the quest for studying this tensor is ON with results of elegance - and the list of workers in this particular field of interest is very long, we have mentioned here only a few of them (some more references on Lanczos potential shall find their place in the subsequent chapters).

The construction of L_{ijk}, for a given spacetime geometry, is equivalent to solving Weyl–Lanczos Eqs. (4.17)/(4.19) with Eqs. (4.12–4.15) as constraints; and as seen from the above discussion that there are several ways of solving Weyl–Lanczos equation but none of them are as straightforward as one would like them to be. Given the Weyl tensor, it is very difficult to construct the Lanczos potential by integrating directly the Eq. (4.19) through tensorial approach. It is therefore worthwhile to look for an efficient method for finding the Lanczos potential; and the following section and remaining chapters are devoted to a detailed study of Lanczos potential using NP and GHP formalisms.

4.3 Weyl–Lanczos Equation and Tetrad Formalism

In this section, we shall find the NP and GHP version of Weyl–Lanczos equations. These results offer corrections to some typographical errors that have occurred in the earlier NP versions of Weyl–Lanczos equations (c.f., [15, 47]).

It is now easy (although lengthy) to see that when the tetrad $Z^a{}_\mu = \{l^a, n^a, m^a, \bar{m}^a\}$, using the definition of differential operators, is projected on Weyl–Lanczos equation we have the following set of five equations. These equations are the NP version of the Weyl–Lanczos Eq. (4.19):

$$\Psi_0 = 2[(-\delta + \bar{\alpha} + 3\beta)L_0 + (D - 3\epsilon + \bar{\epsilon})L_4 - \bar{\pi}L_0 - 3\sigma L_1 - \bar{\rho}L_4 + 3\kappa L_5] \tag{4.32}$$

$$\Psi_1 = (-\triangle + 3\gamma + \bar{\gamma})L_0 + (-\delta + \bar{\alpha} + \beta)L_1 + (\bar{\delta} - 3\alpha + \bar{\beta})L_4 + (D + \bar{\epsilon} - \epsilon)L_5$$

$$+ (\mu - \bar{\mu})L_0 - (\tau + \bar{\pi})L_1 - 2\sigma L_2 - (\bar{\tau} + \pi)L_4 + (3\rho - \bar{\rho})L_5 + 2\kappa L_6 \tag{4.33}$$

$$\Psi_2 = (-\triangle + \gamma + \bar{\gamma})L_1 + (-\delta + \bar{\alpha} - \beta)L_2 + (\bar{\delta} - \alpha + \bar{\beta})L_5 + (D + \bar{\epsilon} + \epsilon)L_6 - \nu L_0$$

$$+ (2\mu - \bar{\mu})L_1 - (2\tau + \bar{\pi})L_2 - \sigma L_3 - \lambda L_4 - (\bar{\tau} + 2\pi)L_5 + (2\rho - \bar{\rho})L_6 + \kappa L_7 \tag{4.34}$$

$$\Psi_3 = (-\triangle - \gamma + \bar{\gamma})L_2 + (-\delta + \bar{\alpha} - 3\beta)L_3 + (\bar{\delta} + \alpha + \bar{\beta})L_6 + (D + \bar{\epsilon} + 3\epsilon)L_7$$

$$+ 2\nu L_1 + (3\mu - \bar{\mu})L_2 - (\tau + \bar{\pi})L_3 - 2\lambda L_5 - (\bar{\tau} + 3\pi)L_6 + (\rho - \bar{\rho})L_7 \tag{4.35}$$

$$\Psi_4 = 2[(-\triangle - 3\gamma + \bar{\gamma})L_3 + (\bar{\delta} + 3\alpha + \bar{\beta})L_7 + 3\nu L_2 - \bar{\mu}L_3 - 3\lambda L_6 - \bar{\tau}L_7] \tag{4.36}$$

where Ψ_i, $i = 0, 1, ..4$ are five complex scalars defined by Eq. (2.24) and

$$L_0 = L_{hij}l^h m^i l^j \ , \quad L_1 = L_{hij}l^h m^i \bar{m}^j$$

$$L_2 = L_{hij}\bar{m}^h n^i l^j \ , \quad L_3 = L_{hij}\bar{m}^h n^i \bar{m}^j$$

$$L_4 = L_{hij}l^h m^i m^j \ , \quad L_5 = L_{hij}l^h m^i n^j \tag{4.37}$$

$$L_6 = L_{hij}\bar{m}^h n^i m^j \ , \quad L_7 = L_{hij}\bar{m}^h n^i n^j$$

are eight complex Lanczos scalars.

Using 18 NP equations [c.f., Eqs. (2.33a)–(2.33r)], it is easy to solve the set of Eqs. (4.32)–(4.36) as 18 NP equations provide a method for integrating Weyl–Lanczos equations (4.19).

For GHP formalism, we have seen that the spin-coefficients κ', σ', etc., defined in Eq. (3.10) are related to the spin-coefficients defined by Newman and Penrose [cf., Eq. (2.11)] as follows:

$$\nu = -\kappa', \ \lambda = -\sigma', \ \mu = -\rho', \ \pi = -\tau', \ \alpha = -\beta', \ \gamma = -\epsilon' \tag{4.38}$$

while the Weyl scalars $\Psi_r, r = 0, 1, 2, 3, 4$ and Lanczos scalars $L_r, r = 0, 1, ..., 7$ defined, respectively, by Eqs. (2.24) and (4.37) take the form

$$\Psi_0 = \Psi_4' , \ \Psi_1 = \Psi_3' , \ \Psi_2 = \Psi_2' , \ \Psi_3 = \Psi_1' , \ \Psi_4 = \Psi_0'$$

$$L_4 = -L_3' \ , \quad L_5 = -L_2' \tag{4.39}$$

$$L_6 = -L_1' \ , \quad L_7 = -L_0'$$

Now, using Eqs. (4.38) and (4.39) in the NP version of Weyl–Lanczos equations [Eqs. (4.32)–(4.36)], we have the following set of three coupled linear differential equations:

$$\Psi_0 = -2\mathcal{P}L_3' - 2\mathcal{D}L_0 + 2\bar{\tau}'L_0 - 6\sigma L_1 - 6\kappa L_2' + 2\bar{\rho}L_3' \tag{4.40}$$

$$\Psi_1 = -\mathcal{P}'L_0 - \mathcal{D}L_1 - \mathcal{P}L_2' - \mathcal{D}'L_3' + (\bar{\rho}' - \bar{\rho})L_0 - (\bar{\tau} + \bar{\tau}')L_1$$

$$- 2\kappa L_1' - 2\sigma L_2 - (3\rho - \bar{\rho})L_2' - (\tau' - \bar{\tau})L_3' \tag{4.41}$$

$$\Psi_2 = -\mathcal{P}'L_1 - \mathcal{P}L_1' - \mathcal{D}L_2 - \mathcal{D}'L_2' - \kappa L_0' + \kappa'L_0 + (\bar{\rho}' - 2\rho')L_1 + \rho L_1'$$

$$+ (\bar{\tau}' - 2\tau)L_2 + (\bar{\tau} - 2\tau' - 2\rho)L_2' - \sigma L_3 - \sigma'L_3' \tag{4.42}$$

Since $\Psi'_0 = \Psi_4$ and $\Psi'_1 = \Psi_3$, the remaining two Weyl–Lanczos equations are the primed version of Eqs. (4.40) and (4.41).

The GHP version of differential gauge condition (4.15) is given by the following set of three equations:

$$\mathcal{P}'L_2 + \mathcal{D}L_3 + \mathcal{D}'L_1' - \mathcal{P}L_0' + 2\kappa'L_1 - (\rho' + \bar{\rho}')L_2 + (\tau - \bar{\tau}')L_3$$

$$+ 2\sigma'L_2' - (\bar{\tau} - 3\tau')L_1' + (\bar{\rho} + \rho)L_0' = 0 \tag{4.43}$$

$$\mathcal{P}'L_0 + \mathcal{D}L_1 + \mathcal{D}'L_3' - \mathcal{P}L_2' - (\rho' + \bar{\rho}')L_0 + (3\tau + \bar{\tau}')L_1 - 2\sigma L_2$$

$$- (\bar{\tau} + \tau')L_3' + (\bar{\rho} + 3\rho)L_2' - 2\kappa L_1' = 0 \tag{4.44}$$

$$\mathcal{P}'L_1 - \mathcal{D}L_2 + \mathcal{D}'L_2' + \mathcal{P}L_1' + \kappa'L_0 - (2\rho' + \bar{\rho}')L_1 + (2\tau - \bar{\tau}')L_2$$

$$- \sigma L_3 + \sigma'L_3' - (\bar{\tau} + 2\tau')L_2' - (\bar{\rho} + 2\rho)L_1' - \kappa L_0' = 0 \tag{4.45}$$

References

1. Acevedo, M., Lopez-Bonilla, J., Sosa-Pedroza, J.: Apeiron **9**, 43–48 (2002)
2. Ahsan, Z.: Tensors: Mathematics of Differential Geometry and Relativity. Second Printing: PHI Learning Pvt. Ltd, Delhi (2017)
3. Ahsan, Z., Bilal, M.: Int. J. Theo. Phys. **49**, 2713–2722 (2010)
4. Ahsan, Z., Bilal, M.: Int. J. Theo. Phys. **50**, 1752–1768 (2011)
5. Ahsan, Z., Bilal, M.: J. Tensor Soc. **6**(2), 127–134 (2012)
6. Ahsan, Z., Bilal, M.: Int. J. Theo. Phys. **52**, 4275–4282 (2013)
7. Ahsan, Z., Ahsan, N., Ali, S.: Bull. Cal. Math. Soc. **93**(5), 407–422 (2001)
8. Ahsan, Z., Ahsan, N., Ali, S.: Maths. Today **19**(2), 25–34 (2001)
9. Ahsan, Z., Caltenco, J.H., Lopez-Bonilla, J.L.: Annalen der Physik **16**(4), 311–313 (2007)
10. Ahsan, Z., Caltencho, J.H., Linares, Y.M.R., Lopez-Bonilla, J.: Commun. Phys. **20**(1), 9–14 (2010)
11. Ahsan, Z., Bilal, M., Lopez-Bonilla, J.: J. Vectorial Relativ. **5**(3), 1–8 (2010)
12. Andersson, F.A., Edgar, S.B.: J. Math. Phys. **41**, 2990–3001 (2000)
13. Aquino, N., Lopez Bonilla, J., Nunez-Yepez, H.N., Salas-Brito, A.L.: Phys. A-Math. Gen. **28**, L375–L379 (1995)
14. Ares de Parga, G., Lopez Bonilla, J., Ovando, G., Matos, T.: Rev. Mex. Fis. **35**(3), 393–409 (1989)
15. Ares de Parga, G., Oscar Chavoya, A., Lopez Bonilla, J.L.: J. Math. Phys. **30**, 1294–1295 (1989)
16. Arreaga, G., Lopez-Bonilla, J., Ovando, G.: Indian J. Pure Appl. Maths. **31**, 85–91 (2000)
17. Bampi, F., Cavigilia, G.: Gen. Rel. Grav. **15**, 375–386 (1983)
18. Bergqvist, G.: J. Math. Phys. **38**, 3142–3154 (1997)
19. Caltenco, J.H., Lopez-Bonilla, J., Morales, J., Ovando, G.: Chinese. J. Phys. **39**, 397–400 (2001)
20. Cartin, D.: The Lanczos potential as a spin-2 field. arXiv:hep-th/0311185v1 (20 Nov. 2003)
21. Dolan, P., Kim, C.W.: Proc. Roy. Soc. Lond. A **447**, 557–575 (1994)
22. Dolan, P., Kim, C.W.: Proc. Roy. Soc. Lond. A **447**, 577–585 (1994)
23. Dolan, P., Muratori, B.D.: J. Math. Phys. **39**, 5404–5420 (1998)
24. Edgar, S.B.: Mod. Phys. Lett. A **9**, 479–482 (1994)

25. Edgar, S.B.: Gen. Rel. Grav. **26**, 329–332 (1994)
26. Edgar, S.B., Höglund, A.: Proc. Roy. Soc. Lond. A **453**, 835–851 (1997)
27. Gaftoi, V., Lopez-Bonilla, J. L., Morales, J., Ovando, G., Peña, J.J.: J. Moscow Phys. Soc. **6**, 267–278 (1996)
28. Gaftoi, V., Lopez-Bonilla, J., Ovando, G.: Int. J. Theo. Phys. **38**, 939–943 (1999)
29. Gaftoi, V., Lopez-Bonilla, J., Ovando, G.: Nuovo Cimento B **113**, 1489–1492 (1998)
30. Gaftoi, V., Lopez Bonilla J.L., Ovando, G., Peña, J.J.: Nuovo Cimento B **113**, 1493–1496 (1998)
31. Illge, R.: Gen. Rel. Grav. **20**, 551–564 (1988)
32. Lanczos, C.: Rev. Mod. Phys. **34**, 379–389 (1962)
33. Lopez-Bonilla, J.L., Morales, J., Naverrete, D., Rosales, M.: Class. Quantum Grav. **10**, 2153–2156 (1993)
34. Lopez-Bonilla, J., Ovando, G., Rivera, J.: Nuovo Cimento B **112**, 1433–1436 (1997)
35. Lopez-Bonilla, J., Ovando, G., Rivera, J.: Indian J. Pure Appl. Maths. **28**, 1355–1360 (1997)
36. Lopez-Bonilla, J., Nunez-Yepez, H.N., Salas-Brito, A.L.: J. Phys. A : Math. Gen **30**, 3663–3669 (1997)
37. Lopez-Bonilla, J., Ovando, G.: Gen. Rel. Grav. **31**, 1931–1934 (1999)
38. Lopez-Bonilla, J., Morales, J., Ovando, G.: Gen. Rel. Grav. **31**, 413–415 (1999)
39. Massa, E., Pagani, E.: Gen. Rel. Grav. **16**, 805–816 (1984)
40. Mena, F.C., Tod, P.: Lanczos potential and a definition of gravitational entropy for perturbed FLRW spacetimes. arXiv:gr-qc/0702057v1 (9 Feb. 2007)
41. Mora, C., Sanchez, R.: Lat. Am. J. Phys. Educ. **1**(1), 78–82 (2007)
42. Novello, M., Velloso, A.L.: Gen. Rel. Grav. **19**, 1251–1265 (1987)
43. Roberts, M.D.: Gen. Rel. Grav. **20**, 775–792 (1989). Mod. Phys. Lett. A **4**, 2739 (1988)
44. Roberts, M.D.: Il Nouvo Cimento **B110**, 1105–1176 (1996)
45. Taub, A.H.: Comp. Math. Appl. **1**, 377 (1975)
46. del Castillo, T.: G. J. Math. Phys. **36**, 195–200 (1995)
47. Zund, J.D.: Ann. Mat. Pure Appl. **104**, 239–268 (1975)

Chapter 5
Lanczos Potential for Algebraically Special Spacetimes

5.1 Introduction

In this chapter, using GHP formalism, we shall first find the Lanczos potential for arbitrary Petrov types II and D spacetimes and then using NP formalism, we shall obtain the Lanczos potential for Petrov types III and N spacetimes. The results thus obtained are supported by examples.

5.2 Lanczos Potential for Petrov Type II Spacetimes

It is very difficult to solve Weyl–Lanczos Eqs. (4.40)–(4.42), but a comparison of these equations with GHP-field Eqs. (3.23a)–(3.23f′) suggests a solution for arbitrary Petrov type II spacetime. We can now choose the null tetrad $\{l^i, n^i, m^i, \bar{m}^i\}$ such that $\Psi_0 = \Psi_1 = 0$ (c.f., [26]). Thus, using Eqs. (4.40)–(4.42) and GHP-field Eqs. (3.23a)–(3.23f′), a possible general solution is given by (see also [4])

$$L_0 = \kappa \quad , \quad L_3' = -\sigma'$$

$$L_1 = \frac{1}{3}\rho \quad , \quad L_2' = -\frac{1}{2}\tau'$$

$$L_2 = -\frac{1}{2}\tau \quad , \quad L_1' = \frac{1}{3}\rho' \tag{5.1}$$

$$L_3 = -\sigma \quad , \quad L_0' = \kappa'$$

Therefore, if L_i $(i = 0, 1, \ldots, 7)$ [from Eq. (5.1)] are known, then from the completeness relation

$$L_{ijk} = M_{ijk} + \bar{M}_{ijk} \tag{5.2}$$

© Springer Nature Singapore Pte Ltd. 2019
Z. Ahsan, *The Potential of Fields in Einstein's Theory of Gravitation*,
https://doi.org/10.1007/978-981-13-8976-4_5

between the Lanczos tensor L_{ijk} and Lanczos scalars L_i we can construct the Lanczos spin tensor which in turn generates the gravitational field (the Weyl tensor) through Weyl–Lanczos Eq. (4.19) where

$$M_{ijk} = L_0 U_{ij} n_k + L_1 (W_{ij} n_k - U_{ij} m_k) + L_2 (V_{ij} n_k - W_{ij} m_k) - L_3 V_{ij} m_k$$
$$+ L_3' U_{ij} \bar{m}_k - L_2' (U_{ij} l_k - W_{ij} \bar{m}_k) - L_1' (W_{ij} l_k - V_{ij} \bar{m}_k) - L_0' V_{ij} l_k \qquad (5.3)$$

and

$$U_{ij} = -n_i \bar{m}_j + n_j \bar{m}_i \ , \quad V_{ij} = l_i m_j - l_j m_i \qquad (5.4)$$
$$W_{ij} = l_i n_j - l_j n_i + m_i \bar{m}_j - m_j \bar{m}_i$$

Moreover, since the spacetime under consideration is of Petrov type II (algebraically special), the Goldberg–Sachs theorem in such case demands that

$$\kappa = \sigma = 0 \qquad (5.5)$$

and thus Eq. (5.1) leads to a simpler form of Lanczos scalars for Petrov type II fields as

$$L_0 = 0 \ , \quad L_3 = 0$$

$$L_1 = \frac{1}{3}\rho \ , \quad L_2' = -\frac{1}{2}\tau'$$

$$L_2 = -\frac{1}{2}\tau \ , \quad L_1' = \frac{1}{3}\rho' \qquad (5.6)$$

$$L_3' = -\sigma' \ , \quad L_0' = \kappa'$$

Example Consider the Robinson–Trautman metrics. They represent the spacetimes which admit a geodetic, hypersurface orthogonal, shear-free and expanding null congruences. Such metrics are the general relativistic analogues of Lienard-Wiechert solutions of Maxwell equations. Moreover, the Robinson–Trautman spacetimes generalize the assumption of spherical symmetry by having topologically equivalent two spheres rather than strictly two spheres. Physically, the Robinson–Trautman metrics can be thought of as representing an isolated gravitationally radiating system. The Robinson–Trautman metric of Petrov type II is [26, 28]

$$ds^2 = -2(U^0 - 2\gamma^0 r - \Psi_2^0 r^{-1})du^2 + 2dudr - \frac{r^2}{2P^2}d\zeta d\bar{\zeta} \qquad (5.7)$$

The derivative operators are

$$D = \frac{\partial}{\partial r}\,,\ \nabla = \frac{\partial}{\partial x^2} + i\frac{\partial}{\partial x^3} = 2\frac{\partial}{\partial \zeta}\,,\ \bar{\nabla} = 2\frac{\partial}{\partial \bar{\zeta}} \tag{5.8}$$

$$\nabla^2 = \nabla\bar{\nabla}\,,\ \ \zeta = x^2 + ix^3,$$

$$\delta = \frac{P}{r}(\delta^k{}_2 + i\delta^k{}_3)\frac{\partial}{\partial x^k}$$

and the spin-coefficients are

$$\kappa = \sigma = \tau = \pi = \lambda = \epsilon = 0\,,\ \ \beta = -\bar{\alpha}\,,\ \rho = \bar{\rho}$$

$$\rho = -r^{-1}\,,\ \ \alpha = \alpha^0 r^{-1}\,,\ \ \beta = -\bar{\alpha}^0 r^{-1} \tag{5.9}$$

$$\gamma = \gamma^0 - \frac{1}{2}\Psi_2^0 r^{-1}\,,\ \ \mu = \mu^0 r - 1 - \Psi_2^0 r^{-2}\,,\ \ \nu = \nu^0 - \Psi_3^0 r$$

with

$$\mu^0 = \bar{\mu}^0 = U^0\,,\ \ \Psi_2^0 = \bar{\Psi}_2^0\,,\ \ U = U^0 - 2\gamma^0 r - \Psi_2 r^{-1}$$

The function $P = P(u, \zeta, \bar{\zeta})$ is real and $^0\xi^k = P(\delta^k{}_2 + i\delta^k{}_3)$. Also,

$$\alpha^0 = \frac{1}{2}P\bar{\nabla}\log P\,,\ \ \gamma^0 = -\frac{\partial}{\partial u}(\log P)$$

$$\nu^0 = -\frac{1}{2}P\nabla\{\frac{\partial}{\partial u}(\log P^2)\}\,,\ \ \mu^0 = -\frac{1}{2}P^2\nabla^2\log P^2$$

The Weyl scalars are

$$\Psi_0 = \Psi_1 = 0\,,\ \ \Psi_2 = \Psi_2^0(u)r^{-3}\,,\ \ \Psi_3 = \Psi_3^0(\zeta)r^{-2} \tag{5.10}$$

$$\Psi_4 = \Psi_4^0 r^{-1} - (P\bar{\nabla}\Psi_3^0 + 2\alpha^0\Psi_3^0)r^{-2}$$

where

$$\Psi_2^0 = \bar{\Psi}_2^0 = \text{constant}$$

$$\Psi_3^0 = P\bar{\nabla}\mu^0\,,\ \ \Psi_4^0 = P\bar{\nabla}\nu^0 + 2\alpha^0\nu^0$$

Now, from Eqs. (5.1) and (5.9), the Lanczos scalars for Robinson–Trautman metric (5.7) are given by

$$L_0 = L_2 = L_3 = L_4 = L_5 = 0$$

$$L_1 = -\frac{1}{3}r^{-1}\,,\ \ L_6 = -L_1' = \frac{1}{3}(\mu^o r^{-1} - \Psi_2^0 r^{-2}) \tag{5.11}$$

$$L_7 = -L_0' = \nu^0 - \Psi_3^0 r$$

It may be noted from Eq. (5.11) that the Lanczos potential for Robinson–Trautman metric of Petrov type II depends upon the radial coordinate r.

5.3   Lanczos Potential for Petrov Type D Spacetimes

It is known that most of the physically significant spacetime solutions are of Petrov type D. Some of the familiar members of this class are Schwarzschild, Reissner-Nördstrom, Kerr, Kerr–Newman, Vaidya and Gödel metrics. Moreover, since GHP formalism has proved useful in the past for studying the Petrov type D gravitational fields [5, 6, 17, 19, 20], it therefore seems worthwhile to study Lanczos potential for such gravitational fields through GHP formalism. In this section, we shall obtain the Lanczos potential for an arbitrary Petrov type D spacetime, which in turn provide a solution to Weyl–Lanczos equations.

A comparison of Eqs. (4.37) and (2.11a) reveals that there is a proportionality between Lanczos scalars and spin-coefficients, for example, $L_0 \propto \kappa$, $L_1 \propto \rho$, $L_2 \propto \tau'$, $L_3 \propto \sigma'$, etc., that is, there is some linear relationship between Lanczos scalars and spin-coefficients. This is precisely the case with some of the important metrics (c.f., [12–15], Sects. 5.4 and 5.5, Chaps. VI and VII of the present text).

Consider an arbitrary vacuum spacetime of Petrov type D with the null tetrad $\{l^i, n^i, m^i, \bar{m}^i\}$ or $\{l^i, l'^i, m^i, \bar{m}^i\}$. Choose l^i and n^i to lie in the direction of the degenerate principal null vectors so that $\Psi_0 = \Psi_1 = \Psi_1' = \Psi_0' = 0$ and $\Psi_2 \neq 0$. Now, using Weyl–Lanczos Eqs. (4.40)–(4.42), the primed version of Eqs. (4.40)–(4.41) and GHP-field Eq. (3.23), a possible general solution is given by (see also [2])

$$L_0 = \kappa \quad , \quad L_4 = -L_3' = -\sigma$$

$$L_1 = \frac{1}{3}\rho \quad , \quad L_5 = -L_2' = -\frac{1}{3}\tau$$

$$L_2 = -\frac{1}{3}\tau' \quad , \quad L_6 = -L_1' = \frac{1}{3}\rho' \tag{5.12}$$

$$L_3 = -\sigma' \quad , \quad L_7 = -L_0' = \kappa'$$

Hence, if Eq. (5.12) are known, then from the completeness relation (5.2) between the Lanczos tensor L_{ijk} and Lanczos scalars L_r $(r = 0, 1, \ldots 7)$, we can construct the Lanczos spin tensor which in turn generates the Weyl tensor (gravitational field) through Weyl–Lanczos Eq. (4.19).

Moreover, as the spacetime under consideration is of Petrov type D, Goldberg–Sachs theorem leads to

$$\kappa = \sigma = \kappa' = \sigma' = 0 \tag{5.13}$$

and thus Eq. (5.13) lead to

$$L_0 = 0 \quad , \quad L_4 = -L_3' = 0$$

$$L_1 = \frac{1}{3}\rho \quad , \quad L_5 = -L_2' = -\frac{1}{3}\tau$$

$$L_2 = -\frac{1}{3}\tau' \quad , \quad L_6 = -L_1' = \frac{1}{3}\rho' \quad\quad (5.14)$$

$$L_3 = 0 \quad , \quad L_7 = -L_0' = 0$$

which is much simpler form of Lanczos scalars for Petrov type D fields.

Example Consider the Kerr spacetime [9] for which the only non-vanishing spin-coefficients are ρ, ρ', τ and τ' and the non-zero component of Weyl scalar is Ψ_2. Since for Kerr spacetime $\kappa = \sigma = \kappa' = \sigma' = 0$; $\Psi_0 = \Psi_1 = \Psi_1' = \Psi_0' = 0$, therefore using Eq. (5.14), GHP-field Eq. (3.23) and GHP Bianchi identities (3.27) along with their primed versions, we have

$$\tau = -\psi\rho\bar{\rho} \quad\quad (5.15)$$

$$\tau' = \psi\rho^2 \quad\quad (5.16)$$

$$\mathcal{P}\tau' = \mathcal{D}'\rho = 2\rho\tau' \qu\quad (5.17)$$

$$\Psi_2 = M\rho^3 \qu\quad (5.18)$$

where M is the mass parameter of the Kerr spacetime and the constant of integration ψ satisfies $\mathcal{P}\psi = 0$ [5, 6, 9].

From Eqs. (5.14), (5.16) and (5.18), the Lanczos potential for the Kerr spacetime is given by

$$L_1 = \frac{1}{3}\left(\frac{\Psi_2}{M}\right)^{\frac{1}{3}}, \quad L_2 = -\frac{\psi}{27}\left(\frac{\Psi_2}{M}\right)^{\frac{2}{3}} \qu\quad (5.19)$$

along with their primed versions. Equation (5.19) clearly shows that the Lanczos potential of the Kerr spacetime is related to the mass parameter of the Kerr spacetime and the Coulomb component of the gravitational field. Moreover, the Lanczos potential of the Kerr black hole depends only on one of the spin-coefficient ρ or τ' [as they are related through Eq. (5.16)] (see also Chap. 7).

5.4 Lanczos Potential for Petrov Type III Spacetimes

It is known that Petrov type III regions are associated with a kind of longitudinal gravitational radiation. In such regions, the tidal forces have a shearing effect. Petrov type III radiation decay is proportional to $\frac{1}{r^2}$. It is for such reasons that this section is devoted to the study of Lanczos potential for Petrov type III gravitational fields.

With a choice of NP-null tetrad $\{m^i, \bar{m}^i, l^i, n^i\}$, the 16 real independent components of the Lanczos potential L_{ijk} are given by the eight complex scalars L_i

$$L_0 = L_{ijk}m^i n^j n^k, L_1 = L_{ijk}m^i n^j \bar{m}^k, L_2 = L_{ijk}l^i \bar{m}^j n^k, L_3 = L_{ijk}l^i \bar{m}^j \bar{m}^k$$

$$L_4 = L_{ijk}m^i n^j m^k, L_5 = L_{ijk}m^i n^j l^k, L_6 = L_{ijk}l^i \bar{m}^j m^k, L_7 = L_{ijk}l^i \bar{m}^j l^k \quad (5.20)$$

which for Petrov type III spacetimes are given by [8]

$$L_0 = -\kappa, \ L_1 = -\rho/3, \ L_2 = \pi/3, \ L_3 = \lambda$$

$$L_4 = -\sigma, \ L_5 = -\tau/3, \ L_6 = \mu/3, \ L_7 = \nu \quad (5.21)$$

Remark It may be noted that for conformally flat spacetime all the components of Lanczos potential are zero.

Using Eq. (5.21), we shall now find the Lanczos potential for some of the known Petrov type III metrics. Here we shall write the null tetrad for each spacetime under consideration along with other necessary information such as non-zero spin-coefficients, intrinsic derivatives, etc. (see also [3]).

5.4.1 Kaigorodov Metric

The Kaigorodov space is a homogeneous Einstein space whose metric in (x, y, v, u) is given by [21]

$$ds^2 = 2(kx)^{-2}(dx^2 + dy^2) - 2du(dv + 2\frac{v}{x}dx) + \frac{4}{3}k^{-1}xdy + 2x^4du \quad (5.22)$$

where k is constant.

The null tetrad for the above line element is

$$l^i = \frac{15}{4}kx\delta_2^i - \frac{257}{8}\sqrt{2}\delta_3^i - \frac{1}{\sqrt{2}x^2}\delta_4^i, \ n^i = -\sqrt{2}x^2\delta_3^i$$

$$m^i = (\frac{1}{2}kx - \frac{15}{4}i\sqrt{2})\delta_1^i - i(\frac{1}{2}kx + \frac{15}{4}\sqrt{2})\delta_2^i - (kv - \frac{59}{6}ix^2)\delta_3^i \quad (5.23)$$

and the intrinsic derivatives are

$$D = \frac{15}{4}kx\frac{\partial}{\partial y} - \frac{257}{8}\sqrt{2}\frac{\partial}{\partial v} - \frac{1}{\sqrt{2}x^2}\frac{\partial}{\partial u} , \quad \Delta = -\sqrt{2}x^2\frac{\partial}{\partial v}$$

$$\delta = (\frac{1}{2}kx - \frac{15}{4}i\sqrt{2})\frac{\partial}{\partial x} - i(\frac{1}{2}kx + \frac{15}{4}\sqrt{2})\frac{\partial}{\partial y} - (kv - \frac{59}{6}ix^2)\frac{\partial}{\partial v} \quad (5.24)$$

so that the non-zero spin-coefficients are

$$\gamma = \frac{ik}{4}\sqrt{2} , \ \mu = \frac{ik}{2}49\sqrt{2} , \ \lambda = \frac{ik}{8}45\sqrt{2}$$

$$\frac{8}{153v} = \frac{\alpha}{2} = \tau = \beta = -\pi = \frac{k}{2} \quad (5.25)$$

The non-zero component of Weyl scalar is

$$\Psi_3 = \frac{1}{2\sqrt{2}}ik^2 \quad (5.26)$$

The Lanczos scalars for the metric (5.22), using Eqs. (5.21) and (5.25) are found to be as

$$L_0 = L_1 = L_4 = 0$$

$$L_2 = -\frac{k}{6} , \ L_3 = \frac{ik}{8}45\sqrt{2} , \ L_5 = -\frac{k}{6} \quad (5.27)$$

$$L_6 = -\frac{ik}{6}49\sqrt{2} , \ L_7 = \frac{153}{16}k$$

These equations show that the Lanczos potential for Kaigorodov metric depends upon the constant k. If $k = 0$ then Eq. (5.26) leads to $\Psi_3 = 0$ which shows that Kaigorodov spacetime is conformally flat; and Eq. (5.27) yields $L_i = 0$, $i = 0, 1, 2, \ldots 7$.

5.4.2 A Type III Solution with Twist

A type III solution with twist is characterized by the following conditions [11]

$$\kappa = \sigma = \alpha = \beta = \gamma = \epsilon = \mu = \pi = 0$$

$$\Psi_0 = \Psi_1 = \Psi_2 = \Psi_4 = 0 \quad (5.28)$$

$$\rho , \lambda , \tau , \nu \neq 0 ; \ \Psi_3 \neq 0 ; \ \rho \neq \bar{\rho}$$

Debney et al. [10] have exhibited a type III twist vacuum solution which is mani-
festly different from Held [18] and Robinson [25]. Moreover, upon a particular spe-
cialization of the equations leading to this solution, Debney et al. [11] also deduced
the Held and Robinson solutions. The outline of this solution is given as follows:

Let $x^i = (u, r, x, y)$ denotes the real coordinate system, where r is radial parame-
ter along the family of null congruence $\Gamma(l^i)$. The functions $U = \bar{U}$, $X^i = \bar{X}^i$, ω, ξ^i
($i = 1, 3, 4$) are components of tetrad vectors and the covariant tetrad 1-form

$$L \equiv l_i dx^i, \quad N \equiv n_i dx^i, \quad M \equiv m_i dx^i, \quad \bar{M} \equiv \bar{m}_i dx^i \qquad (5.29)$$

eventually leads to the metric

$$ds^2 = g_{ij} dx^i dx^j = LN + NL - M\bar{M} - \bar{M}M$$

The solution of NP field equations [c.f., Eq. (2.33)] leads to

$$\tau = \tau^\circ \rho, \quad \lambda = \lambda^\circ \rho, \quad \nu = \nu^\circ \rho$$

$$\omega = \omega^\circ \bar{\rho}, \quad \xi^i = \xi^{\circ i} \bar{\rho}, \quad \Psi_3 = \Psi_3^\circ \rho^2 \qquad (5.30)$$

$$U = U^\circ + \omega^\circ \bar{\tau} + \bar{\omega}^\circ \tau, \quad X^i = X^{\circ i} + \xi^{\circ i} \bar{\tau} + \bar{\xi}^{\circ i} \tau$$

where the 'degree sign' as a superscript indicates that function is independent of r
and

$$\rho = \frac{-1}{r + i\Sigma} \quad, \text{where } \Sigma(u, x, y) \neq 0 \qquad (5.31)$$

assures that the null ray congruence $\Gamma(l^i)$ having the tangent vector l^i has non-
vanishing twist. Thus, the covariant tetrad 1-form [cf., Eq. (5.29)] becomes

$$L = (X^{\circ 1})^{-1} [du - \text{Re}(\xi^{\circ 1}) dx - \text{Im}(\xi^{\circ 1}) dy]$$

$$N = dr - \text{Re}(\omega^\circ) dx - \text{Im}(\omega^\circ) dy - U^\circ L \qquad (5.32)$$

$$M = \frac{-1}{2\rho} [dx + i dy] + \tau^\circ L$$

with

$$\xi^{\circ 3} = 1, \ \xi^{\circ 4} = i, \ X^{\circ 3} = X^{\circ 4} = 0, \ \xi^{\circ 1} = i\Omega$$

$$\lambda^\circ = -\frac{3}{4}\frac{1}{x^2} = U^\circ, \ \tau^\circ = -\frac{3}{2}\frac{1}{x}, \ \Psi_3^\circ = -\frac{3}{2}\frac{1}{x^3} \qquad (5.33)$$

$$\nu^\circ = \lambda^\circ \tau^\circ + \Psi_3^\circ, \ X^{\circ 1} = x^{-\frac{3}{2}}, \ \omega^\circ = i(2\tau^\circ \Sigma - \mathcal{D}\Sigma)$$

$$\mathcal{D} \equiv i\Omega\frac{\partial}{\partial u} + \frac{\partial}{\partial x} + i\frac{\partial}{\partial y}$$

where the real functions $\Omega(x, y, z)$ must be determined.

Using Eqs. (5.30) and (5.33), the non-zero spin-coefficients are given by

$$\lambda = -\frac{3}{4}\frac{\rho}{x^2}, \quad \nu = -\frac{3}{8}\frac{\rho}{x^3}, \quad \tau = -\frac{3}{2}\frac{\rho}{x} \tag{5.34}$$

while the non-zero Weyl scalar is

$$\Psi_3 = -\frac{3}{2}\frac{\rho^2}{x^3} \tag{5.35}$$

Now using Eqs. (5.21), (5.27), (5.30) and (5.33), the Lanczos scalars for the type III solution with twist are given by

$$L_0 = L_2 = L_4 = L_6 = 0$$

$$L_1 = -\frac{1}{3}\rho, \quad L_3 = -\frac{3}{4}\frac{\rho}{x^2} \tag{5.36}$$

$$L_5 = \frac{1}{2x}\rho, \quad L_7 = \frac{-3}{8}\frac{\rho}{x^3}$$

which shows that the Lanczos potential depends only on one spin-coefficient ρ.

5.5 Lanczos Potential for Petrov Type N Spacetimes

The most interesting, but rather difficult and little explored, solutions of Einstein vacuum field equations are the Petrov type N solutions [1]. From the physical point of view, they represent spacetimes filled entirely with gravitational radiation. Moreover, it is known from the peeling-off theorem that the dominant term in an asymptotic expansion of the gravitational field is of Petrov type N. Thus, such fields can be considered as containing the main features of the gravitational radiation far from the sources. Due to the importance of Petrov type N fields, this section is devoted to the study of Lanczos potential for such spacetimes.

A possible solution of Weyl–Lanczos equations for Petrov type N fields is given by [8]

$$L_0 = -\kappa/2, \quad L_1 = -\rho/6, \quad L_2 = \pi/6, \quad L_3 = \lambda/2$$

$$L_4 = -\sigma/2, \quad L_5 = -\tau/6, \quad L_6 = \mu/6, \quad L_7 = \nu/2 \tag{5.37}$$

Using Eq. (5.37), we shall now obtain the Lanczos potential of some well-known type N solutions. In each case, we shall write the null tetrad for the metric under consideration along with other necessary information such as non-zero spin-coefficients, intrinsic derivatives, etc. (see also [7]).

5.5.1   Generalized pp-Waves

A class of solution to Einstein–Maxwell equations for the null electrovac Petrov type N gravitational field has been given by Radhakrishna and Singh [24]. The metric is

$$ds^2 = -2U du^2 + 2du dr - \frac{1}{2} \bar{P}[(dx^3)^2 + (dx^4)^2] \tag{5.38}$$

where $U = U(u, x^3, x^4)$ and $P = \bar{P}(x^3, x^4)$ satisfies the partial differential equation

$$P P_{(,z\bar{z})} - P_{(,z)} P_{(,\bar{z})} = 0 \tag{5.39}$$

The derivative with respect to $z = x^3 + ix^4$ is denoted by

$$\partial_z = \frac{1}{2}(\partial_3 - i\partial_4) \tag{5.40}$$

It may be noted that for $P = \frac{1}{2}$ and $U = -H(u, x^3, x^4)$, the solution (5.38) reduces to the plane-fronted gravitational waves solution given by Kundt [22]. The null vectors chosen are

$$l^i = \delta_2^i , \quad n^i = \delta_1^i + U\delta_2^i , \quad m^i = P(\delta_3^i + i\delta_4^i) \tag{5.41}$$

The intrinsic derivatives are

$$D = \frac{\partial}{\partial r} , \quad \Delta = \frac{\partial}{\partial u} + U\frac{\partial}{\partial r} , \quad \delta = P(\frac{\partial}{\partial x^3} + i\frac{\partial}{\partial x^4}) \tag{5.42}$$

and the non-vanishing spin-coefficients are

$$\alpha = -\bar{\beta} = P_{,z} , \quad \nu = -2PU_{,z} \tag{5.43}$$

while the non-zero components of Ricci, Weyl and Maxwell scalars, respectively, are

$$\Phi_{22} = -4P^2 U_{,z\bar{z}} , \quad \Psi_4 = -4(P^2 U_{,z})_{,z} , \quad \phi_2 = PQ(u, z) \tag{5.44}$$

From Eqs. (5.43) and (5.37), the Lanczos scalars for the metric (5.38) are

$$L_7 = PU_{,z} = -\frac{1}{2}\nu , \quad L_i = 0 , \quad i = 0, 1, 2, \ldots 6 \tag{5.45}$$

which shows that the Lanczos potential for the metric (5.38) depends only on one
spin-coefficient ν.

5.5.2 Pure Radiation Metric

A solution of Einstein field equations corresponding to pure radiation fields has been
obtained by Ludwig and Edgar [23]. The line element in (u, r, x, y) coordinates is

$$ds^2 = (-2xW^\circ - \frac{8r^2}{x^2}\tilde{P}^2)du^2 + 2dudr - \frac{4r}{x}dudx - \frac{1}{8\tilde{P}^2}(dx^2 + dy^2) \quad (5.46)$$

where $\tilde{P}$ is an arbitrary non-zero constant and W° is an arbitrary function of the three
non-radial coordinates. A degree sign denotes that the quantity is independent of r.
A suitable tetrad for the metric (5.46) is

$$l^i = \delta_2^i , \quad n^i = \delta_1^i + (xW^\circ - \frac{12r^2}{x^2}\tilde{P}^2)\delta_2^i - \frac{16r}{x}\tilde{P}^2\delta_3^i ,$$

$$m^i = 2\tilde{P}(-\delta_3^i + i\delta_4^i) \quad (5.47)$$

so that the intrinsic derivatives are

$$D = \frac{\partial}{\partial r} , \quad \Delta = \frac{\partial}{\partial u} + (xW^\circ - \frac{12r^2}{x^2}\tilde{P}^2)\frac{\partial}{\partial r} - \frac{16r}{x}\tilde{P}^2\frac{\partial}{\partial x} ,$$

$$\delta = 2\tilde{P}(-\frac{\partial}{\partial x} + i\frac{\partial}{\partial y}) \quad (5.48)$$

The non-zero spin-coefficients for the metric (5.46) are given by

$$2\alpha = 2\beta = \pi = \tau = \frac{2\tilde{P}}{x}$$

$$\gamma = \frac{12r}{x}\tilde{P}^2 , \quad \lambda = \mu = \frac{8r}{x^2}\tilde{P}^2 \quad (5.49)$$

$$\nu = 2\tilde{P}[\frac{\partial(xW^\circ)}{\partial x} + i\frac{\partial(xW^\circ)}{\partial y}] + \frac{48r^2\tilde{P}^3}{x^3}$$

while the non-zero components of the Weyl and Ricci scalars, respectively, are

$$\Psi_4 = -4x\tilde{P}^2\left[\frac{\partial^2 W^\circ}{\partial x^2} - \frac{\partial^2 W^\circ}{\partial y^2} + 2i\frac{\partial^2 W^\circ}{\partial x\partial y}\right]$$

$$\Phi_{22} == -4x\tilde{P}^2[\frac{\partial^2(xW^\circ)}{\partial x^2} + \frac{\partial^2(xW^\circ)}{\partial y^2}] \tag{5.50}$$

Using Eqs. (5.37) and (5.49), the Lanczos scalars for the metric (5.46) are given by

$$L_0 = L_1 = L_4 = 0$$

$$L_2 = \frac{\tilde{P}}{3x}, \quad L_3 = \frac{4r}{x^2}\tilde{P}^2, \quad L_5 = -\frac{\tilde{P}}{3x} \tag{5.51}$$

$$L_6 = \frac{4r}{3x^2}\tilde{P}^2, \quad L_7 = \tilde{P}[\frac{\partial(xW^\circ)}{\partial x} + i\frac{\partial(xW^\circ)}{\partial y}] + \frac{24r^2\tilde{P}^3}{x^3}$$

5.5.3 Kaigorodov Metric

The Kaigorodov space is a homogeneous Einstein space and it describes a pp-waves (plane-fronted waves with parallel rays) propagating in anti-de Sitter space. The metric in (x, y, v, u) coordinates is given by [21]

$$ds^2 = 2(ax)^{-2}(dx^2 + dy^2) - 2du(dv + 2vx^{-1}dx + xdu) \tag{5.52}$$

where $a \neq 0$ is a constant. The null tetrad for the line element (5.52) is

$$l^i = -\sqrt{x}\delta_3^i + \frac{1}{\sqrt{x}}\delta_4^i, \quad n^i = \sqrt{x}\delta_3^i,$$

$$m^i = a(\frac{x}{2}\delta_1^i + ix\delta_2^i - v\delta_3^i) \tag{5.53}$$

and the intrinsic derivatives are

$$D = -\sqrt{x}\frac{\partial}{\partial v} + \frac{1}{\sqrt{x}}\frac{\partial}{\partial u}, \quad \Delta = \sqrt{x}\frac{\partial}{\partial v}, \tag{5.54}$$

$$\delta = a(\frac{x}{2}\frac{\partial}{\partial x} + ix\frac{\partial}{\partial y} - v\frac{\partial}{\partial v})$$

so that the non-zero spin-coefficients are

$$\frac{8}{5}\alpha = 4\beta = \tau = -\pi = \frac{\nu}{3} = \frac{a}{2} \tag{5.55}$$

and the Weyl scalar is

$$\Psi_4 = \frac{3}{2}a^2 \tag{5.56}$$

The Lanczos scalars for the metric (5.52) are found to be as [using Eqs. (5.55) and (5.37)]

$$L_0 = L_1 = L_3 = L_4 = L_6 = 0$$

$$L_2 = L_5 = -\frac{a}{12}, \ L_7 = \frac{3}{4}a \tag{5.57}$$

which shows that the Lanczos potential for Kaigorodov solution depends upon the non-zero constant a. If $a = 0$, then from Eq. (5.56), $\Psi_4 = 0$ and Kaigorodov spacetime becomes conformally flat; and also from Eq. (5.57) $L_i = 0, i = 0, 1, 2, \ldots, 7$ (see also Sect. 5.4.1).

5.5.4 Twist-Free Gravitational Field

The Hauser solution [16, 27] is characterized by the following conditions:

$$\kappa = \sigma = \alpha = \beta = \gamma = \epsilon = \mu = \pi = 0$$

$$\Psi_0 = \Psi_1 = \Psi_2 = \Psi_3 = 0, \ \Lambda = \Phi_{mn} = 0 \ (m, n = 0, 1, 2) \tag{5.58}$$

$$\rho = \lambda = \tau = \nu \neq 0, \quad \Psi_4 \neq 0, \quad \rho \neq \bar{\rho}$$

A spacetime satisfying Eq. (5.58) is said to be Hauser-like. It is twist-free when $\rho = \bar{\rho}$. Wilkes and Zund [27] have obtained a Hauser-like type N twist-free solution. The outline of this solution is as follows:

Let $x^i = (u, r, x, y)$ denote the local coordinates, where r is an affine radial parameter along the family of null congruence $\Gamma(l^i)$. The metric tensor is given by

$$g^{ij} = l^i n^j + n^i l^j - m^i \bar{m}^j - \bar{m}^i m^j \tag{5.59}$$

where the tetrad vectors are chosen as

$$l^i = \delta_2^i, \ n^i = X^1 \delta_1^i + U \delta_2^i + X^3 \delta_3^i + X^4 \delta_4^i$$

$$m^i = \xi^1 \delta_1^i + \omega \delta_2^i + \xi^3 \delta_3^i + \xi^4 \delta_4^i \tag{5.60}$$

The intrinsic derivatives are

$$D = \frac{\partial}{\partial r}, \ \Delta = X^1 \frac{\partial}{\partial u} + U \frac{\partial}{\partial r} + X^3 \frac{\partial}{\partial x} + X^4 \frac{\partial}{\partial y}$$

$$\delta = \xi^1 \frac{\partial}{\partial u} + \omega \frac{\partial}{\partial r} + \xi^3 \frac{\partial}{\partial x} + \xi^4 \frac{\partial}{\partial y} \tag{5.61}$$

While the NP field equations lead to

$$\rho = -\frac{1}{r} , \ \lambda = \lambda^\circ \rho , \ \tau = \tau^\circ \rho , \ \nu = \nu^\circ \rho \ ; \ \nu^\circ = \lambda^\circ \tau^\circ , \ \xi^i = \xi^{\circ i}$$

$$X^i = X^{\circ i} + \bar{\xi}^{\circ i} \tau + \xi^{\circ i} \bar{\tau} , \quad (i = 1, 3, 4) \tag{5.62a}$$

$$\omega = 0 , \ \Psi_4 = \Psi_4^\circ \rho$$

with

$$\xi^{\circ 3} = 1 , \ \xi^{\circ 4} = i , \ X^{\circ 3} = X^{\circ 4} = 0 , \ \xi^{\circ 1} = 0$$

$$\lambda^\circ = U^\circ = -g^2 , \ \tau^\circ = g , \ X^{\circ 1} = e^\Omega , \ \Psi_4^\circ = 2g\dot{g}e^\Omega \tag{5.62b}$$

where $g(u)$ is an arbitrary non-constant function, a dot denotes the differentiation with respect to u and a degree sign as a superscript denotes a function of r. From Eqs. (5.62a) and (5.62b), the non-vanishing spin-coefficients are

$$\rho = -\frac{1}{r} , \ \lambda = \frac{1}{r}g^2 , \ \tau = -\frac{1}{r}g , \ \nu = \frac{1}{r}g^3 \tag{5.63}$$

Thus, from Eqs. (5.63) and (5.37), the Lanczos scalars for the Hauser-like twist-free solution are given by

$$L_0 = L_2 = L_4 = L_6 = 0$$

$$L_1 = \frac{1}{6r} , \ L_3 = \frac{1}{2r}g^2 , \ L_5 = \frac{1}{6r}g , \ L_7 = \frac{1}{2r}g^3 \tag{5.64}$$

which shows that the Lanczos potential depends upon the affine radial parameter r.

References

1. Ahsan, Z.: Indian J. Pure Appl. Maths. **31**(2), 215–225 (2000)
2. Ahsan, Z., Bilal, M.: Int. J. Theo. Phys. **49**, 2713–2722 (2010)
3. Ahsan, Z., Bilal, M.: J. Tensor Soc. **6**(2), 127–134 (2012)
4. Ahsan, Z., Bilal, M.: Int. J. Theo. Phys. **52**, 4275–4282 (2013)
5. Ahsan, Z., Ahsan, N., Ali, S.: Bull. Cal. Math. Soc. **93**(5), 407–422 (2001)
6. Ahsan, Z., Ahsan, N., Ali, S.: Maths. Today **19**(2), 25–34 (2001)
7. Ahsan, Z., Bilal, M., Lopez-Bonilla, J.: J. Vectorial Relat. **5**(3), 1–8 (2010)
8. Ares de Parga, G., Oscar Chavoya, A., Lopez Bonilla, J.L.: J. Math. Phys. **30**, 1294–1295 (1989)
9. Bergqvist, G.: J. Math. Phys. **38**, 3142–3154 (1997)
10. Debney, G.C., Wilkes, J.M., Zund, J.D.: Tensor. N. S. **35**, 267 (1981)
11. Debney, G.C., Wilkes, J.M., Zund, J.D.: Tensor. N. S. **37**, 90 (1982)

12. Gaftoi, V., Lopez-Bonilla, J.L., Morales, J., Naverrete, D., Ovando, G.: Rev. Mex. Fis. **36**, 498 (1990); **37**, 638 (1991)
13. Gaftoi, V., Lopez-Bonilla, J.L., Morales, J., Ovando, G., Peña, J.J.: J. Moscow Phys. Soc. **6**, 267–278 (1996)
14. Gaftoi, V., Morales, J., Ovando, G., Peña, J.J. Nuovo Cimento B **113**, 1297 (1998)
15. Gaftoi, V., Lopez Bonilla J.L., Ovando, G., Peña, J.J.: Nuovo Cimento B **113**, 1493–1496 (1998)
16. Hauser, I.: Phys. Rev. Lett. **33**, 1112 (1974)
17. Held, A.: Comm. Math. Phys. **37**, 311 (1974)
18. Held, A.: Lett. al Nuovo Cimento **11**, 545 (1974)
19. Held, A.: J. Math. Phys. **17**, 39–45 (1976)
20. Held, A.: Gen. Rel. Grav. **7**, 177 (1976)
21. Kaigorodov, V.R.: Sov. Phys. Doklady **7**, 893 (1963)
22. Kundt, W.: Z. Phyzik **163**, 77 (1961)
23. Ludwig, G., Edgar, S.B.: Class Quantum Grav. **14**, 3453 (1997)
24. Radhakrishna, L., Singh, N.I.: J. Math. Phys. **25**, 2293 (1984)
25. Robinson, I.: Gen. Rel. Grav. **6**, 423 (1975)
26. Stephani, H., Kramer, D., Maccallum, M., Hoenselaers, C., Herlt, E.: Exact Solutions of Einstein's Field Equations, 2nd edn. Cambridge University Press, UK (2003)
27. Wilkes, J.M., Zund, J.D.: Tensor N. S. **37**, 16 (1982)
28. Zund, J.D.: Ann. Mat. Pure Appl. **104**, 239–268 (1975)

Chapter 6
Lanczos Potential and Perfect Fluid Spacetimes

6.1 Introduction

The construction of Lanczos potential, for a given spacetime geometry, is equivalent to solving Eq. (4.19) with Eqs. (4.12–4.15) as constraints. In this chapter, NP formalism has been used to obtain the Lanczos potential and Lanczos scalars for perfect fluid spacetimes, which in turn leads to a solution of Weyl–Lanczos equations (see also [3]). The method of general observers has been considered and the NP versions of the kinematical quantities and the equations satisfied by them have been obtained in Sect. 6.2. While in Sect. 6.3, the Lanczos potentials for shear-free, irrotational perfect fluid and Bianchi type I spacetimes have been obtained. As an example to Bianchi type I models, the Lanczos potential for Kasner metric has been calculated, and it has been observed that the Lanczos scalars for the Kasner spacetime depend upon time and the constant appearing in the Kasner metric. The Lanczos potential for Gödel spacetime has also been obtained and it is found that the Lanczos scalars depend upon the parameter which is responsible for the vorticity (rotation) of the fluid and thus a possible physical meaning is assigned to Lanczos potential.

6.2 General Observers and NP Formalism

The basic covariant variables for a gravitational field with perfect fluid source are:

(A) the fluid scalars—θ (expansion), $\tilde{\rho}$ (energy density) and p (pressure);
(B) the fluid spatial vectors—$\dot{u}_i$ (4-acceleration), ω_i (vorticity);
(C) the spatial trace-free symmetric tensors—σ_{ij} (fluid shear), the electric (E_{ij}) and the magnetic (H_{ij}) parts of the Weyl tensor;
(D) the projection tensor h_{ij} which projects orthogonal to the fluid 4-velocity vector u_i.

© Springer Nature Singapore Pte Ltd. 2019
Z. Ahsan, *The Potential of Fields in Einstein's Theory of Gravitation*,
https://doi.org/10.1007/978-981-13-8976-4_6

These quantities for a unit time-like vector u_i are given as follows [2, 6, 7] {the time-like vector field u_i is often taken as the 4-velocity of the fluid}:
(i) The *expansion scalar* θ is defined as

$$\theta = u^i_{;i} \tag{6.1}$$

(ii) The *acceleration vector* a_i is given by

$$a_i = \dot{u}_i = u_{i;j} u^j \tag{6.2a}$$

and is such that

$$a_i u^i = 0 \tag{6.2b}$$

(iii) The *projection tensor* h_{ij} is defined as

$$h_{ij} = g_{ij} - u_i u_j \tag{6.3a}$$

and has the properties as

$$h_{ij} = h_{ji} \,, \; h^k_i h_{kj} = h_{ij} \,, \; h^i_i = 3 \,, \; h_{ij} u^j = 0 \tag{6.3b}$$

(iv) The symmetric *shear tensor* σ_{ij} is defined by

$$\sigma_{ij} = h^k_i h^l_j u_{(k;l)} - \frac{1}{3}\theta h_{ij} \tag{6.4a}$$

and has the properties that

$$\sigma_{ij} u^j = 0 \,, \; \sigma^i_i = 0 \tag{6.4b}$$

(v) The skew-symmetric *vorticity* or *rotation tensor* ω_{ij} is given by

$$\omega_{ij} = h^k_i h^l_j u_{[k;l]} \tag{6.5a}$$

This tensor satisfies the equation

$$\omega_{ij} u^i = \omega_{ij} u^j = 0 \tag{6.5b}$$

and is equivalent to a *vorticity vector*

$$\omega^i = \frac{1}{2}\eta^{ijkl}\omega_{jk} u_l \tag{6.5c}$$

so that

$$\omega_{ij} = \eta_{ijkl}\omega^k u^l \tag{6.5d}$$

where η^{ijkl} is completely skew-symmetric Levi-Civita tensor.

From Eqs. (6.2b), (6.4b) and (6.5b), it may be noted that the kinematical quantities a_i, σ_{ij} and ω_{ij} are space-like.

(vi) The *electric* and *magnetic* parts of the Weyl tensor as measured by an observer with a time-like 4-velocity vector u^i are defined, respectively, as (c.f., [1, 2])

$$E_{ij} = C_{ijkl}u^k u^l \tag{6.6}$$

$$H_{ij} = {}^* C_{ijkl}u^k u^l = \frac{1}{2}\eta_{ik}{}^{mn}C_{mnjl}u^k u^l \tag{6.7}$$

and satisfy the following:

$$E_{ij} = E_{ji} \, , \ E_{ij}u^j = 0 \, , \ E_{ij}g^{ij} = E_k^k = 0 \tag{6.8}$$

$$H_{ij} = H_{ji} \, , \ H_{ij}u^j = 0 \, , \ H_{ij}g^{ij} = H_k^k = 0 \tag{6.9}$$

That is, both E_{ij} and H_{ij} are symmetric, space-like and traceless.

The Weyl tensor C_{ijkl} is said to be purely electric if $H_{ij} = 0$ and purely magnetic if $E_{ij} = 0$, and in terms of E_{ij} and H_{ij} the Weyl tensor can be decomposed as [2]

$$C_{hijk} = 2u_h u_j E_{ik} + 2u_i u_k E_{hj} - 2u_h u_k E_{ij} - 2u_i u_j E_{hk} + g_{hk} E_{ij} + g_{ij} E_{hk} - g_{hj} E_{ik} - g_{ik} E_{hj}$$

$$+ \eta_{hi}{}^{pq} u_p u_k H_{qj} - \eta_{hi}{}^{pq} u_p u_j H_{qk} + \eta_{jk}{}^{pq} u_i u_p H_{hq} - \eta_{jk}{}^{pq} u_h u_p H_{iq} \tag{6.10}$$

(vii) The covariant derivative of u_i may be decomposed into its irreducible parts as

$$u_{i;j} = \sigma_{ij} + \frac{1}{3}\theta h_{ij} + \omega_{ij} + a_i u_j \tag{6.11}$$

(For a simple derivation of this equation, see [2]).

(viii) The energy density $\tilde{\rho}$ and the pressure p are related through the energy–momentum tensor T_{ij} of the perfect fluid as

$$T_{ij} = \tilde{\rho} u_i u_j - p h_{ij} \tag{6.12}$$

so that the relativistic equations for the conservation of energy and momentum are

$$T^{ij}{}_{;j} = 0 \tag{6.13}$$

We shall now derive the NP versions of these kinematical quantities and the equations satisfied by them. These results will be used to obtain the Lanczos potential for the perfect fluid spacetimes (see also [3]).

Let $Z^i{}_\mu = \{l^i, n^i, m^i, \bar{m}^i\}$ be the complex null tetrad, where l^i, n^i are real vectors and $m^i, \bar{m}^i$ are the complex null vectors. All the inner products between the tetrad vectors vanish except $l^i n_i = 1 = -m^i \bar{m}^i$.

The metric tensor g_{ij}, in terms of the tetrad vectors, can be written as

$$g_{ij} = l_i n_j + n_i l_j - m_i \bar{m}_j - \bar{m}_i m_j \tag{6.14}$$

We choose the 4-velocity vector u^i as

$$u^i = \frac{1}{\sqrt{2}}(l^i + n^i) \tag{6.15a}$$

such that

$$u_i u^i = 1 \tag{6.15b}$$

From Eqs. (6.14) and (6.15), the projection tensor h_{ij} [defined by Eq. (6.3a)] takes the form

$$h_{ij} = l_{(i} n_{j)} - 2m_{(i} \bar{m}_{j)} - \frac{1}{2}(l_i l_j + n_i n_j) \tag{6.16}$$

while from Eqs. (6.15a), (2.41) and (2.42), we have

$$u_{i;j} = \tfrac{1}{\sqrt{2}}\{(\gamma + \bar{\gamma})l_j l_i - \bar{\tau}l_j m_i - \tau l_j \bar{m}_i + (\epsilon + \bar{\epsilon})n_j l_i - \bar{\kappa}n_j m_i - \kappa n_j \bar{m}_i$$

$$-(\alpha + \bar{\beta})m_j l_i + \bar{\sigma}m_j m_i + \rho m_j \bar{m}_i - (\bar{\alpha} + \beta)\bar{m}_j l_i + \bar{\rho}\bar{m}_j m_i + \sigma \bar{m}_j \bar{m}_i$$

$$-(\gamma + \bar{\gamma})l_j n_i + \nu l_j m_i + \bar{\nu}l_j \bar{m}_i - (\epsilon + \bar{\epsilon})n_j n_i + \pi n_j m_i + \bar{\pi}n_j \bar{m}_i$$

$$+(\alpha + \bar{\beta})m_j n_i - \lambda m_j m_i - \bar{\mu}m_j \bar{m}_i + (\bar{\alpha} + \beta)\bar{m}_j n_i - \mu \bar{m}_j m_i - \bar{\lambda}\bar{m}_j \bar{m}_i\} \tag{6.17}$$

so that

$$u_{(i;j)} = \tfrac{1}{\sqrt{2}}\{(\gamma + \bar{\gamma})l_{(j} l_{i)} + (\epsilon + \bar{\epsilon} - \gamma - \bar{\gamma})n_{(j} l_{i)} + (\epsilon + \bar{\epsilon})n_{(j} n_{i)}$$

$$-(\bar{\tau} + \alpha + \bar{\beta} - \nu)l_{(j} m_{i)} - (\bar{\kappa} - \pi - \alpha - \bar{\beta})n_{(j} m_{i)}$$

$$+(\bar{\sigma} - \lambda)m_{(j} m_{i)} + (\rho + \bar{\rho} - \mu - \bar{\mu})m_{(j} \bar{m}_{i)} \tag{6.18a}$$

$$-(\tau + \bar{\alpha} + \beta - \bar{\nu})l_{(j} \bar{m}_{i)} - (\kappa - \bar{\pi} - \bar{\alpha} - \beta)n_{(j} \bar{m}_{i)}$$

$$+(\sigma - \bar{\lambda})\bar{m}_{(j} \bar{m}_{i)}\}$$

and

$$u_{[i;j]} = \frac{1}{\sqrt{2}}\{(\epsilon + \bar{\epsilon} + \gamma + \bar{\gamma})n_{[j}l_{i]} + (\alpha + \bar{\beta} - \bar{\tau} + \nu)l_{[j}m_{i]}$$

$$+(\bar{\alpha} + \beta - \tau + \bar{\nu})l_{[j}\bar{m}_{i]} - (\bar{\kappa} - \pi + \alpha + \bar{\beta})n_{[j}m_{i]} \qquad (6.18b)$$

$$-(\kappa - \bar{\pi} + \bar{\alpha} + \beta)n_{[j}\bar{m}_{i]} + (\rho - \bar{\rho} + \mu - \bar{\mu})m_{[j}\bar{m}_{i]}\}$$

Also, from Eqs. (2.44), (2.45) and (6.15a), the expansion θ [given by Eq. (6.1)] can be expressed in terms of spin-coefficients as

$$\theta = \frac{1}{\sqrt{2}}(\epsilon + \bar{\epsilon} - \gamma - \bar{\gamma} - \rho - \bar{\rho} + \mu + \bar{\mu}) \qquad (6.19)$$

Thus, the fluid is expansion-free if and only if

$$\epsilon + \bar{\epsilon} = \gamma + \bar{\gamma} = \rho + \bar{\rho} = \mu + \bar{\mu} = 0 \qquad (6.20)$$

Using Eqs. (6.2a) and (6.17), the acceleration vector a_i can be expressed in terms of spin-coefficients as

$$a_i = \frac{1}{\sqrt{2}}(\epsilon + \bar{\epsilon} + \gamma + \bar{\gamma})v_i + \frac{1}{2}\{(\nu - \pi - \bar{\kappa} - \bar{\tau})m_i + (\bar{\nu} - \bar{\pi} - \kappa - \tau)\bar{m}_i\}$$
$$(6.21)$$

where the vector v_i is defined by

$$v_i = \frac{1}{\sqrt{2}}(l^i - n^i) \qquad (6.22a)$$

and is such that

$$v_i v^i = -1 , \quad v_i u^i = 0 \qquad (6.22b)$$

Thus, the fluid is geodetic (or acceleration-free) if and only if

$$\epsilon + \bar{\epsilon} = \gamma + \bar{\gamma} = 0 \qquad (6.23a)$$

$$\nu - \pi - \bar{\kappa} - \bar{\tau} = 0 \qquad (6.23b)$$

Moreover, from Eqs. (6.16), (6.18a) and (6.19), the shear tensor σ_{ij} can be expressed in terms of spin-coefficients as

$$\sigma_{ij} = \frac{1}{3\sqrt{2}}\{2(\epsilon + \bar{\epsilon}) + \rho + \bar{\rho} - 2(\gamma + \bar{\gamma}) - \mu - \bar{\mu}\}(v_i v_j - \tfrac{1}{2}m_i \bar{m}_j - \tfrac{1}{2}\bar{m}_i m_j)$$

$$+ \tfrac{1}{4}\{[2(\alpha + \bar{\beta}) + \pi + \bar{\tau} - \nu - \bar{\kappa}](v_i m_j + v_j m_i)\}$$

$$+ \tfrac{1}{4}\{[2(\bar{\alpha} + \beta) + \bar{\pi} + \tau - \bar{\nu} - \kappa](v_i \bar{m}_j + v_j \bar{m}_i)\}$$

$$+ \tfrac{1}{2\sqrt{2}}\{(\lambda - \bar{\sigma})m_i m_j + (\bar{\lambda} - \sigma)\bar{m}_i \bar{m}_j\}$$

$$\text{(6.24)}$$

Thus, the fluid is shear-free if and only if

$$\epsilon + \bar{\epsilon} = \gamma + \bar{\gamma} = \rho + \bar{\rho} = \mu + \bar{\mu} = 0 \tag{6.25a}$$

$$\pi + \bar{\tau} - \nu - \bar{\kappa} = 0 \tag{6.25b}$$

$$\alpha + \bar{\beta} = 0 , \quad \lambda - \bar{\sigma} = 0 \tag{6.25c}$$

While, using Eqs. (6.16) and (6.18b), the rotation tensor ω_{ij} can be expressed in terms of spin-coefficients as

$$\omega_{ij} = \tfrac{1}{4}\{[2(\alpha + \bar{\beta}) + \nu + \bar{\kappa} - \pi - \bar{\tau}](v_i m_j - v_j m_i)\}$$

$$+ \tfrac{1}{4}\{[2(\bar{\alpha} + \beta) + \bar{\nu} + \kappa - \bar{\pi} - \tau](v_i \bar{m}_j - v_j \bar{m}_i)\} \tag{6.26}$$

$$+ \tfrac{1}{2\sqrt{2}}\{\rho - \bar{\rho} + \mu - \bar{\mu}\}(m_i \bar{m}_j - \bar{m}_i m_j)$$

Thus, the fluid is irrotational ($\omega_{ij} = 0$) if and only if

$$\alpha + \bar{\beta} = 0 \tag{6.27a}$$

$$\nu + \bar{\kappa} - \pi - \bar{\tau} = 0 \tag{6.27b}$$

$$\rho = \bar{\rho} , \quad \mu = \bar{\mu} \tag{6.27c}$$

Following relations shall also be useful:

$$u_i l^i = u_i n^i = -v_i l^i = v_i n^i = \frac{1}{\sqrt{2}} \tag{6.28a}$$

$$u_i m^i = u_i \bar{m}^i = v_i m^i = v_i \bar{m}^i = 0 \tag{6.28b}$$

$$h_{ij} l^i l^j = -\frac{1}{2} , \; h_{ij} n^i l^j = \frac{1}{2} , \; h_{ij} n^i n^j = -\frac{1}{2} , \; h_{ij} m^i \bar{m}^j = -1 ,$$

$$h_{ij} m^i l^j = h_{ij} l^i \bar{m}^j = h_{ij} \bar{m}^i l^j = h_{ij} \bar{m}^i \bar{m}^j = h_{ij} n^i \bar{m}^j = h_{ij} m^i m^j = 0 \tag{6.29}$$

6.3 Perfect Fluid Spacetimes and Lanczos Potential

When a relativistic model of the universe is investigated then the matter contents of the universe are taken to be as described by the energy–momentum tensor of a perfect fluid. The Einstein field equations with a perfect fluid source are given by

$$R_{ij} - \frac{1}{2}Rg_{ij} + \Lambda g_{ij} = -(\tilde{\rho}u_i u_j - ph_{ij}) \tag{6.30}$$

where u_i ($u_i u^i = 1$) is the velocity of the fluid, $\tilde{\rho} > 0$ the energy density, p the pressure and Λ the cosmological constant.

The investigation of the exact solutions of Eq. (6.30) is of interest due to the following reasons:
(i) as cosmological models,
(ii) as interior solutions which have to be matched with vacuum exterior solution,
(iii) as representing the propagation of gravitational waves in matter.

6.3.1 Shear-Free and Irrotational Spacetimes

Under different assumptions, a large number of researchers have found the exact solutions of Eq. (6.30) (cf., [14]). For example, one can obtain the Friedmann–Lemaitre–Robertson–Walker (FLRW) model with the assumption of homogeneity and isotropy, while the conformal flatness assumption of the spacetime either leads to an FLRW model or a non-homogeneous cosmological model, provided that there exists a relation between density and pressure or not. Moreover, with the assumptions of spherical symmetry and geodesic motion, one can get a homogeneous FLRW metric. All such static solutions either lead to de Sitter spacetime or to a Schwarzschild interior solution. All of these solutions are shear-free and irrotational and it has been shown by Maiti [12] that an irrotational and shear-free perfect fluid spacetime is either spherically symmetric or pseudo-spherical or plane symmetric. It therefore seems worthwhile to find the Lanczos potential for shear-free and vorticity-free (irrotational) perfect fluid spacetimes. The Newman–Penrose formalism has been used for such an investigation.

The Lanczos potential (up to a gauge) for irrotational and shear-free spacetime is given by [13]

$$L_{ijk} = a_i u_j u_k - a_j u_i u_k \tag{6.31}$$

Now, using Eqs. (6.21), (6.25) and (6.27) in Eq. (6.31), we have the following theorem.

Theorem 6.1 *For a shear-free irrotational perfect fluid spacetime, the Lanczos potential (up to a gauge) is*

$$L_{ijk} = 2(\pi - \bar{\kappa})m_{[i}u_{j]}u_k + 2(\bar{\pi} - \kappa)\bar{m}_{[i}u_{j]}u_k \qquad (6.32)$$

and the Lanczos scalars, using Eqs. (6.28) and (6.32), are

$$L_1 = L_3 = L_4 = L_6 = 0, \quad L_0 = L_5 = -\bar{L}_2 = -\bar{L}_7 = \frac{1}{2}(\bar{\pi} - \kappa) \qquad (6.33)$$

If the algebraic gauge conditions [c.f., Eq. (4.15a)] are used, then Eq. (6.31) leads to

$$L_{ijk} = a_j\left\{\frac{2}{3}u_iu_k + \frac{1}{3}h_{ik}\right\} - a_i\left\{\frac{2}{3}u_ju_k + \frac{1}{3}h_{jk}\right\} \qquad (6.34)$$

Thus, using Eq. (6.21), (6.25) and (6.27) in Eq. (6.34), we have the following theorem.

Theorem 6.2 *The Lanczos potential for a perfect fluid spacetime without shear and rotation is given by*

$$L_{ijk} = 2(\pi - \bar{\kappa})\left\{\frac{2}{3}m_{[j}u_{i]}u_k + \frac{1}{3}m_{[j}h_{i]k}\right\} + 2(\bar{\pi} - \kappa)\left\{\frac{2}{3}\bar{m}_{[j}u_{i]}u_k + \frac{1}{3}\bar{m}_{[j}h_{i]k}\right\}$$
$$(6.35)$$

[provided that Eq. (4.15) are satisfied] and the Lanczos scalars [using Eqs. (6.28), (6.29) and (6.35)] are

$$L_0 = -\bar{L}_2 = \frac{1}{3}L_5 = -\bar{L}_7 = -\frac{1}{6}(\bar{\pi} - \kappa), \quad L_1 = L_3 = L_4 = L_6 = 0 \qquad (6.36)$$

Remark If the Lanczos scalars are known, then from the completeness relation

$$L_{ijk} = M_{ijk} + \bar{M}_{ijk} \qquad (6.37)$$

between the Lanczos tensor L_{ijk} and Lanczos scalars L_r ($r = 0, 1, \ldots 7$) we can construct the Lanczos spin tensor which in turn generates the Weyl tensor (gravitational field) through Weyl–Lanczos Eq. (4.19) where

$$M_{ijk} = L_0 U_{ij}n_k + L_1(W_{ij}n_k - U_{ij}m_k) + L_2(V_{ij}n_k - W_{ij}m_k) - L_3 V_{ij}m_k$$

$$- L_4 U_{ij}\bar{m}_k + L_5(U_{ij}l_k - W_{ij}\bar{m}_k) + L_6(W_{ij}l_k - V_{ij}\bar{m}_k) + L_7 V_{ij}l_k \qquad (6.38)$$

and

$$U_{ij} = -n_i\bar{m}_j + n_j\bar{m}_i , \quad V_{ij} = l_im_j - l_jm_i,$$

$$W_{ij} = l_in_j - l_jn_i + m_i\bar{m}_j - m_j\bar{m}_i \qquad (6.39)$$

6.3.2 Bianchi Type I Universes

The present universe, on a large scale, is assumed to be isotropic and homogeneous. Physical cosmology is based on relativistic Friedmann–Lemaitre–Robertson–Walker (FLRW) models which describe the universe as completely homogeneous and isotropic in all its evolution.

In recent years cosmological models, which are not FLRW, have also been studied. This study is motivated by two reasons—(i) the need to investigate the consequence of general relativity (the theory in which all cosmological models are described) and (ii) because the FLRW cannot explain several features and riddles of the present universe. For instance, why the present universe is FLRW while it might have been anisotropic at early times. Also, the perturbed FLRW models lead to decaying models, which would have been much more important in the past, suggesting a finite deviation from FLRW at early epochs.

The study of non-FLRW models begins with the anisotropic and spatially homogeneous models in which a three-dimensional isometry group G_3 acts simply transitively on the hypersurface of homogeneity. These models are included in the Bianchi classification and the simplest expanding anisotropic models are Bianchi Type I family where the three-dimensional spatial space is the rest space of matter. The metric can be expressed in matter co-moving coordinates as [9]

$$ds^2 = -dt^2 + X^2(t)dx^2 + Y^2(t)dy^2 + Z^2(t)dz^2 , \ u^i = \delta^i{}_4 \qquad (6.40)$$

This is the simplest generalization of spatially flat FLRW models to allow for different expansion factors in three orthogonal directions; the corresponding average expansion scale factor is $S(t) = (XYZ)^{\frac{1}{3}}$ [6]. These models are spatially homogeneous, and the fluid flow (orthogonal to homogeneous surfaces) is necessarily geodetic and irrotational. Thus, these models follow the restriction

$$\dot{u}_i = \omega_{ij} = 0 \qquad (6.41)$$

and Eq. (6.11) for such models reduces to

$$u_{i;j} = \sigma_{ij} + \frac{1}{3}\theta h_{ij} \qquad (6.42)$$

The Lanczos potential in such cases is (see also [5])

$$L_{ijk} = -\frac{2}{3}u_{[i}\sigma_{j]k} \qquad (6.43)$$

Thus, using Eqs. (6.23), (6.24) and (6.27) in Eq. (6.43), we have the following theorem.

Theorem 6.3 *The Lanczos potential for Bianchi Type I perfect fluid spacetime is given by*

$$L_{ijk} = \frac{8}{9\sqrt{2}}(\rho - \bar{\mu})\left\{2u_{[j}v_{i]}v_k - u_{[j}m_{i]}\bar{m}_k - u_{[j}\bar{m}_{i]}m_k\right\}$$

$$+\frac{4}{3}(\pi - \bar{\kappa})\left\{u_{[j}v_{i]}m_k + u_{[j}m_{i]}v_k\right\}$$

$$+\frac{4}{3}(\bar{\pi} - \kappa)\left\{u_{[j}v_{i]}\bar{m}_k + u_{[j}\bar{m}_{i]}v_k\right\} \qquad (6.44)$$

$$+\frac{4}{3\sqrt{2}}\left\{(\lambda - \bar{\sigma})u_{[j}m_{i]}m_k + (\bar{\lambda} - \sigma)u_{[j}\bar{m}_{i]}\bar{m}_k\right\}$$

and the Lanczos scalars are

$$L_o = -\bar{L}_2 = -L_5 = \bar{L}_7 = -\frac{1}{3}(\bar{\pi} - \kappa),$$

$$L_1 = -L_6 = \frac{2}{9}(\rho - \bar{\mu}), \qquad (6.45)$$

$$L_3 = -\bar{L}_4 = \frac{1}{3}(\lambda - \bar{\sigma})$$

Example: Kasner Metric

It is known that for Bianchi Type I spacetimes, the shear $\sigma^2 \neq 0$. It does not matter how small it is at present, but it will dominate the very early evolution of the model of the universe. This will then approximate to Kasner vacuum solution [10] which is useful in understanding the behaviour of general cosmological singularities. The Kasner metric is given by

$$ds^2 = dt^2 - t^{2p_1}dx^2 - t^{2p_2}dy^2 - t^{2p_3}dz^2 \qquad (6.46)$$

where the constants p_i satisfy the condition

$$p_1 + p_2 + p_3 = 1 = p_1^2 + p_2^2 + p_3^2 \qquad (6.47)$$

If we arrange p_i in ascending order, the conditions (6.47) imply that [11]

$$-\frac{1}{3} \leq p_1 \leq 0 \leq p_2 \leq \frac{2}{3} \leq p_3 \leq 1 \qquad (6.48)$$

so that the model contracts in one direction, while expanding in the other two. The total volume expands is proportional to t. Since there is no physical way to distinguish the coordinates x, y, z (a model that contracts in the y-direction is the same as the model that contracts in x-direction), any permutation of p_i satisfying (6.47) describes the same model.

It may be noted that for $p_1 = 1$, $p_2 = p_3 = 0$, the Kasner solution (6.46) reduces to flat spacetime. The non-flat plane symmetric case ($p_2 = p_3 = \frac{2}{3}$) was given by

Taub [15]. When $p_1 = p_2 = \frac{2}{3}$ and $p_3 = -\frac{1}{3}$ then Kasner metric reduces to an LRS (locally rotationally symmetric) space with Petrov type D metric [14].

Consider the time-like vector field u^a as

$$u^a = \delta_4^a$$

The null tetrad (in NP formalism) for the metric (6.46) is

$$l^a = \frac{1}{\sqrt{2}}(t^{-p_1}, 0, 0, 1)$$

$$n^a = \frac{1}{\sqrt{2}}(-t^{-p_1}, 0, 0, 1) \tag{6.49}$$

$$m^a = \frac{1}{\sqrt{2}}(0, t^{-p_2}, -it^{-p_3}, 0)$$

and the non-zero spin-coefficients are

$$\rho = -\mu = -\frac{(p_3 + p_2)}{2\sqrt{2}t}$$

$$\lambda = -\sigma = -\frac{(p_3 - p_2)}{2\sqrt{2}} \tag{6.50}$$

$$\gamma = -\epsilon = -\frac{p_1}{2\sqrt{2}t}$$

Therefore, from Eq. (6.45) (c.f., Theorem 6.3), the Lanczos scalars for Kasner metric (6.46) with (6.47) are

$$L_0 = L_2 = L_5 = L_7 = 0$$

$$L_1 = -L_6 = -\frac{\sqrt{2}}{9t}(1 - p_1) \tag{6.51}$$

$$L_3 = -L_4 = -\frac{1}{3\sqrt{2}t}(p_3 - p_2)$$

Special Cases:

(i) If we choose the constant p_i as $p_1 = 1$, $p_2 = p_3 = 0$, Eq. (6.51) leads to $L_i = 0$ ($i = 0, 1, 2, \ldots 7$) and the Kasner spacetime reduces to conformally flat spacetime.

(ii) When $p_2 = p_3 = \frac{2}{3}$, then Eq. (6.51) leads to

$$L_0 = L_2 = L_3 = L_4 = L_5 = L_7 = 0$$

$$L_1 = L_6 = -\frac{\sqrt{2}}{9t}(1 - p_1)$$

(iii) If $p_1 = p_2 = \frac{2}{3}$ and $p_3 = -\frac{1}{3}$, then Lanczos scalars given by Eq. (6.51) reduces to

$$L_1 = -L_6 = -\frac{\sqrt{2}}{27t}, \quad L_3 = -L_4 = \frac{1}{3\sqrt{2}t}, \quad L_0 = L_2 = L_5 = L_7 = 0$$

This shows that the Lanczos scalars L_i, $i = 1, 3, 4, 6$ for a locally rotationally symmetric space with Petrov type D metric depend on time t only and is inversely proportional to t.

6.3.3   Gödel Spacetime

The Gödel spacetime, which represents a homogeneous and non-isotropic universe, is the singularity-free solution of Einstein field equations (with non-zero cosmological constant) with source as dust. The metric is [8]

$$ds^2 = dt^2 - dx^2 - dy^2 + \frac{1}{2}e^{2qy}dz^2 + 2e^{qy}dzdt \tag{6.52}$$

where q is a parameter related to the vorticity of the fluid.

The Gödel solution (6.52) is not the realistic model of the universe in which we live (as the model does not allow Hubble expansion), but it does possess a number of strange properties. The solution contains closed time-like curves which would allow for a form of time travel, i.e. an observer can influence his past. The Weyl tensor of the Gödel solution is of Petrov type D [4]. This means that for an appropriately chosen observer, the tidal forces have Coulomb component. In terms of the kinematical quantities, the Gödel solution is characterized by [4, 7]

$$a_i = \theta = \sigma_{ij} = 0, \tag{6.53}$$

$$\omega_{i;j} = 0 \tag{6.54}$$

Equation (6.54) shows that the vorticity vector ω_i is covariantly constant. The Lanczos potential for a perfect fluid spacetime satisfying (6.53) and (6.54) is given by [13]

$$L_{ijk} = \frac{2}{9}[\omega_{ij}u_k + \frac{1}{2}\omega_{ik}u_j - \frac{1}{2}\omega_{jk}u_i] \tag{6.55}$$

Using Eqs. (6.20), (6.23), (6.25) and (6.26), the NP version of (6.55) is given by the following theorem.

Theorem 6.4 *For a spacetime in which there is a field of observers* $u^a = \delta_4^a$ *which is expansion-free, geodetic, shear-free and the vorticity vector is covariantly constant, the Lanczos potential is given by*

$$L_{ijk} = \frac{2}{9\sqrt{2}}(\rho - \bar{\mu})\{2m_{[i}\bar{m}_{j]}u_k + m_{[i}\bar{m}_{k]}u_j - m_{[j}\bar{m}_{k]}u_i\} \tag{6.56}$$

and the Lanczos scalars are

$$L_0 = L_2 = L_3 = L_4 = L_5 = L_7 = 0, \quad L_1 = 6L_6 = \frac{1}{3}(\rho + \mu) \tag{6.57}$$

The null tetrad for the line element (6.52) is

$$l_i = \frac{1}{\sqrt{2}}(-1, 0, e^{qy}, 1)$$

$$n_i = \frac{1}{\sqrt{2}}(1, 0, e^{qy}, 1) \tag{6.58a}$$

$$m_i = \frac{1}{\sqrt{2}}(0, -1, \frac{i}{\sqrt{2}}e^{qy}, 0)$$

and

$$l^i = \frac{1}{\sqrt{2}}(1, 0, 0, 1)$$

$$n^i = \frac{1}{\sqrt{2}}(-1, 0, 0, 1) \tag{6.58b}$$

$$m^i = \frac{1}{\sqrt{2}}(0, 1, -i\sqrt{2}e^{-qy}, i\sqrt{2})$$

such that

$$l_i n^i = 1 = -m_i \bar{m}^i$$

and all other inner products vanish. The non-zero spin-coefficients are

$$\alpha = -\frac{1}{2\sqrt{2}}q, \ \beta = \frac{1}{2\sqrt{2}}q, \ \gamma = -\frac{i}{4}q$$

$$\epsilon = -\frac{i}{4}q, \ \rho = -\frac{i}{2}q, \ \mu = -\frac{i}{2}q \tag{6.59}$$

Thus, using Theorem 6.4, the Lanczos scalars for Gödel spacetime are given by

$$L_1 = -\frac{i}{3}q \ , \ L_6 = -\frac{i}{18}q \tag{6.60}$$

which shows that the Lanczos potential of the Gödel solution is related to the vorticity parameter of the fluid. Thus, a possible physical meaning can be assigned to Lanczos potential.

It may be noted that the only non-vanishing Weyl scalar is

$$\Psi_2 = -\frac{q^2}{6}$$

which shows that the Coulomb component of the gravitational field depends upon the rotation parameter q and can be regarded as radiation due to rotation.

References

1. Ahsan, Z.: Indian J. Pure Appl. Maths. **30**, 863–869 (1999)
2. Ahsan, Z.: Tensors: Mathematics of Differential Geometry and Relativity. PHI Learning Pvt. Ltd., Delhi (2015)
3. Ahsan, Z., Bilal, M.: Int. J. Theo. Phys. **50**, 1752–1768 (2011)
4. Ahsan, Z., Ahsan, N., Ali, S.: Bull. Cal. Math. Soc. **93**(5), 407–422 (2001)
5. Dolan, P., Muratori, B.D.: J. Math. Phys. **39**, 5404–5420 (1998)
6. Ellis, G.F.R.: Relativistic cosmology-article in general relativity and cosmology. In: Sachs, R.K. (Eds.), Proceedings International School of Physics "Enrico Fermi", pp. 104–179. Acad. Press, New York (1971)
7. Ellis, G.F.R., Elst, H.V.: Cosmological Models. arXiv:gr-qc/9812046v5 (2 Sep. 2008)
8. Gödel, K.: Rev. Mod. Phys. **21**, 447–450 (1949)
9. Heckmann, O., Shücking, E.: Article in gravitation-an introduction to current research. In: Witten, L. (Ed.) Wiley New York (1962)
10. Kasner, E.: Am. J. Math. **43**, 217 (1921)
11. Landau, L.D., Lifshitz, E.M.: The Classical Theory of Fields. Pergamon Press, New York, Fourth revised English edition (1975)
12. Maiti, S. R.: Gen. Rel. Grav. **16**(3), 297 (1984)
13. Novello, M., Velloso, A.L.: Gen. Rel. Grav. **19**, 1251–1265 (1987)
14. Stephani, H., Kramer, D., Maccallum, M., Hoenselaers, C., Herlt, E.: Exact Solutions of Einstein's Field Equations, 2nd edn. Cambridge University Press, UK (2003)
15. Taub, A.H.: Ann. Math. **53**, 472 (1951)

Chapter 7
Lanczos Potential For the Spacetime Solutions

7.1 Introduction

In Chaps. 4–6, we have seen that the tetrad formalisms provide an efficient way of finding the Lanczos potentials in some general situations. Using those prescriptions, in this chapter, we shall find the Lanczos potential for some well-known solutions of Einstein and Einstein–Maxwell equations; Sect. 7.2 of this chapter deals with such study. Section 7.3 deals with the situations that what will happen to Lanczos potential when another choice of tetrad is made. The results related to such choice of tetrad are discussed here in detail here along with number of examples. Apart from tetrad formalisms, there are also other methods for finding the Lanczos potential, and in the remaining parts of this chapters, three methods have been discussed and applied to find the Lanczos potential for the Gödel spacetime.

7.2 Some Special Metrics

Here, we shall find the Lanczos potential for some well-known metrics using the techniques of previous chapters and then for the sake of completeness we write down the Lanczos potential obtained by the other workers. In all such cases we write down the null tetrad for the metric under consideration, its non-vanishing spin-coefficients, intrinsic derivatives, etc., and then the Lanczos scalars. Once the Lanczos scalars L_i, $(i = 0, 1, 2, \ldots, 7)$ are known then from the completeness relations (5.2) the Lanczos potential can be obtained for the metric under consideration. We have

© Springer Nature Singapore Pte Ltd. 2019
Z. Ahsan, *The Potential of Fields in Einstein's Theory of Gravitation*,
https://doi.org/10.1007/978-981-13-8976-4_7

7.2.1 Charged Rotating Black Hole

Newman et al. [38] have obtained a solution of Einstein–Maxwell equation which, in (r, θ, ϕ, u) coordinates, is described by the metric [22]

$$ds^2 = [1 - A^{-2}(2mr - e^2)]du^2 + 2dudr + 2aA^{-2}(2mr - e^2)\sin^2\theta dud\phi$$

$$- 2a\sin^2\theta drd\phi - A^2d\theta^2 - [(r^2 + a^2)^2 - Ba^2\sin^2\theta]A^{-2}\sin^2\theta d\phi^2 \qquad (7.1)$$

where $A^2 = r^2 + a^2\cos^2\theta$ and $B = r^2 - 2mr + a^2 + e^2$. The solution (7.1) is now commonly known as Kerr–Newman solution and represents the exterior gravitational field of a charged rotating mass; it contains three parameters—m (mass), e (charge) and a (angular momentum per unit mass). This solution defines a black hole with an event horizon only when $a^2 + e^2 \leq m^2$.

It is known that rotating black holes are formed due to gravitational collapse of a massive spinning star or from the collapse of a collection of stars or gas with a total non-zero angular momentum. Since most of the stars rotate, it is expected that most of the black holes in nature are rotating and thus Kerr–Newman solution represents the gravitational field outside a charged rotating black hole (see also [4]).

The null tetrad vectors for the metric (7.1) are

$$l^i = \delta_2^i, \quad n^i = \frac{1}{A^2}\left[(r^2 + a^2)\delta_1^i - \frac{B}{2}\delta_2^i + a\delta_4^i\right],$$

$$m^i = \frac{1}{C\sqrt{2}}\left[ia\sin\theta\delta_1^i + \delta_3^i + i\csc\theta\delta_4^i\right] \qquad (7.2)$$

where $C = r + ia\cos\theta$. The non-zero spin-coefficients are given by

$$\rho = -\bar{C}^{-1}, \quad \mu = -\frac{B}{2\bar{C}A^2}, \quad \alpha = \frac{2ia - C\cos\theta}{2\sqrt{2}(\bar{C}\bar{C}\sin\theta)}, \quad \beta = \frac{\cot\theta}{2\sqrt{2}C}$$

$$\tau = -\frac{ia\sin\theta}{\sqrt{2}A^2}, \quad \pi = \frac{ia\sin\theta}{\sqrt{2}C\bar{C}}, \quad \gamma = \frac{(r - m)\bar{C} - B}{2\bar{C}A^2} \qquad (7.3)$$

while the non-zero components of Weyl and Maxwell scalars, respectively, are

$$\Psi_2 = -\frac{m}{\bar{C}^3} + \frac{e^2}{C\bar{C}^3} \quad \text{and} \quad \phi_1 = \frac{e}{\sqrt{2}C\bar{C}} \qquad (7.4)$$

which shows that Kerr–Newman solution (7.1) is of Petrov type D with non-null electromagnetic field.

Therefore, from Eqs. (5.14) and (7.3), the Lanczos scalars for the charged rotating black hole are given by

$$L_0 = L_3 = L_4 = L_7 = 0$$

$$L_1 = -\frac{1}{3\bar{C}}, \ L_2 = \frac{ia\sin\theta}{3\sqrt{2}C\bar{C}}, \ L_5 = \frac{ia\sin\theta}{3\sqrt{2}C}, \ L_6 = \frac{B}{6\bar{C}A^2} \quad (7.5)$$

7.2.2 Kerr Black Hole

When $e = 0$ and $a \neq 0$, Eq. (7.1) reduces to the line element of Kerr solution (a rotating black hole) which can be expressed as

$$ds^2 = \left[1 - \frac{2mr}{r^2 + a^2\cos\theta}\right]du^2 + 2dudr + \left[\frac{4amr\sin^2\theta}{r^2 + a^2\cos\theta}\right]dud\phi - 2a\sin^2\theta drd\phi$$

$$- (r^2 + a^2\cos^2\theta)d\theta^2 - \left[\frac{(r^2 + a^2)^2 - (r^2 - 2mr + a^2)a^2\sin^2\theta}{r^2 + a^2\cos^2\theta}\right]\sin^2\theta d\phi^2 \quad (7.6)$$

The non-zero spin-coefficients are given by

$$\rho = -\bar{C}^{-1}, \ \mu = -\frac{2mr - r^2 - a^2}{2\bar{C}A^2}, \ \alpha = \frac{2ia - C\cos\theta}{2\sqrt{2}(\bar{C}\bar{C}\sin\theta)}, \ \beta = \frac{\cot\theta}{2\sqrt{2}C}$$

$$\tau = -\frac{ia\sin\theta}{\sqrt{2}A^2}, \ \pi = \frac{ia\sin\theta}{\sqrt{2}C\bar{C}}, \ \gamma = \frac{(r - m)\bar{C} - (r^2 - 2mr + a^2)}{2\bar{C}A^2} \quad (7.7)$$

where $C = r + ia\cos\theta$ and $A^2 = r^2 + a^2\cos\theta$.

The non-zero components of Weyl scalar are

$$\Psi_2 = -\frac{m}{\bar{C}^3} \quad (7.8)$$

which shows that Kerr metric (7.6) is of Petrov type D. Hence, using the scheme (5.14), the Lanczos scalar for Kerr black hole is given by

$$L_0 = L_3 = L_4 = L_7 = 0$$

$$L_1 = -\frac{1}{3}\rho, \ L_2 = \frac{1}{3}\pi, \ L_5 = -\frac{1}{3}\tau, \ L_6 = -\frac{1}{3}\mu \quad (7.9)$$

where the spin-coefficients ρ, π, τ and μ are given by Eq. (7.7).

It may be noted that Petrov type D fields have only Coulomb component of the gravitational field and thus the Lanczos scalars, given by Eq. (7.9), can act as the potential of the gravitational field of Kerr black hole (in analogy with the electromagnetism).

7.2.3 Reissner–Nördstrom Metric

If the angular momentum a is zero, then the line element (7.1) reduces to Reissner–Nördstrom metric given by

$$ds^2 = \left(1 - \frac{2m}{r} + \frac{e^2}{r^2}\right)du^2 + 2du\,dr - r^2(d\theta^2 + \sin^2\theta\,d\phi^2) \tag{7.10}$$

The null tetrad for the metric (7.10) can be taken as [22]

$$l_i = \delta_2^i, \ n^i = \delta_1^i - \frac{1}{2}\left(1 - \frac{2m}{r} + \frac{e^2}{r^2}\right)\delta_2^i, \ m^i = \frac{1}{r\sqrt{2}}\left(\delta_3^i + i\csc\theta\,\delta_4^i\right) \tag{7.11}$$

so that the non-zero spin-coefficients are

$$\rho = -\frac{1}{r}, \ \mu = -\frac{1}{2r}\left(1 - \frac{2m}{r} + \frac{e^2}{r^2}\right),$$

$$\alpha = -\beta = \frac{1}{2\sqrt{2}r}\cot\theta, \ \gamma = \frac{1}{2r^3}(mr - e^2) \tag{7.12a}$$

and the non-zero components of Weyl and Maxwell scalars are, respectively, given by

$$\Psi_2 = \frac{e^2}{r^4} - \frac{m}{r^3}, \ \phi_1 = \frac{e}{r^2\sqrt{2}} \tag{7.12b}$$

Hence following the prescription (5.14), the Lanczos scalars for the Reissner–Nördstrom metric are

$$L_0 = L_2 = L_3 = L_4 = L_5 = L_7 = 0$$

$$L_1 = \frac{1}{3r}, \ L_6 = -\frac{1}{6r}\left(1 - \frac{2m}{r} + \frac{e^2}{r}\right) \tag{7.13}$$

which clearly indicates that the Lanczos potential for Reissner–Nördstrom solution depends upon the radial coordinate r.

7.2.4 Schwarzschild Exterior Solution

Consider the Schwarzschild metric as

$$ds^2 = -\left(1 - \frac{2m}{r}\right)^{-1}dr^2 - r^2(d\theta^2 + \sin^2\theta^2 d\phi^2) + \left(1 - \frac{2m}{r}\right)dt^2 \tag{7.14}$$

Using Kinnersley null tetrad [9, 27]

$$l^i = \frac{1}{r^2 - 2mr}(r^2, \ r^2 - 2mr, \ 0, \ 0), \ n^i = \frac{1}{2r^2}(r^2, \ 2mr - r^2, \ 0, \ 0),$$

$$m^i = \frac{1}{r\sqrt{2}}(0, \ 0, \ , \ 1, \ i \csc \theta) \tag{7.15}$$

the non-vanishing spin-coefficients are given by

$$\rho = -\frac{1}{r}, \ \mu = -\frac{1}{2r}\left(1 - \frac{2m}{r}\right), \ \alpha = -\beta = -\frac{1}{2\sqrt{2}r} \cot \theta, \ \gamma = \frac{m}{2r^2} \tag{7.16}$$

while the non-zero components of Weyl are

$$\Psi_2 = -\frac{m}{r^3} \tag{7.17}$$

The intrinsic derivatives used here are given by

$$D = l^i \nabla_i = \frac{r^2}{A}\frac{\partial}{\partial t} + \frac{\partial}{\partial r}, \ \Delta = n^i \nabla_i = \frac{1}{2}\frac{\partial}{\partial t} - \frac{A}{2r^2}\frac{\partial}{\partial r},$$

$$\delta = m^i \nabla_i = \frac{1}{r\sqrt{2}}\left(\frac{\partial}{\partial \theta} + i \csc\theta \frac{\partial}{\partial \phi}\right)$$

where $A = r^2 - 2mr$.

Since the Schwarzschild exterior solution is of Petrov type D [cf., Eq. (7.17)], using Eq. (5.14), the Lanczos scalars are

$$L_0 = L_2 = L_3 = L_4 = L_5 = L_7 = 0, \ L_1 = \frac{1}{3}\rho, \ L_6 = -\frac{1}{6}\mu \tag{7.18}$$

where ρ and μ are given by Eq. (7.16).

Consider now the Schwarzschild solution in null coordinates $x^i = (u, r, \theta, \phi)$ as

$$ds^2 = \left(1 - \frac{2m}{r}\right)du^2 + 2dudr - r^2(d\theta^2 + \sin^2 \theta d\phi^2) \tag{7.19}$$

and taking the intrinsic derivatives as [49]

$$D = \frac{\partial}{\partial r}, \ \Delta = \frac{\partial}{\partial u} - \frac{1}{2}\left(1 - \frac{2m}{r}\right)\frac{\partial}{\partial r}, \ \delta = \frac{1}{r\sqrt{2}}\left(\frac{\partial}{\partial \theta} + i \csc\theta \frac{\partial}{\partial \phi}\right) \tag{7.20}$$

the non-zero spin-coefficients are given by

$$\rho = -\frac{1}{r}, \ \mu = -\frac{1}{2r}, \ \alpha = \frac{\alpha^\circ}{r}, \ \beta = -\frac{\bar{\alpha}^\circ}{r}, \ \gamma = -\frac{1}{2r} + \frac{m}{r^2} \qquad (7.21)$$

where $\alpha^\circ = \bar{\alpha}^\circ = -\frac{\cot\theta}{\sqrt{2}}$. Thus, from Eq. (5.14), the Lanczos scalars for Schwarzschild solution in null coordinates are given by

$$L_0 = L_2 = L_3 = L_4 = L_5 = L_7 = 0, \ 2L_6 = -L_1 = \frac{1}{3r} \qquad (7.22)$$

Remark It may be noted from Eq. (7.18) that the Lanczos scalars depend upon two parameters r (the radial coordinate) and m (the mass), while Eq. (7.22) shows that the Lanczos scalars depend only on radial coordinate r. This clearly indicates the non-uniqueness character of Lanczos potential, and this feature of Lanczos potential is in close analogy with that of the potential for electromagnetic field. Moreover, Eq. (7.22) tells us that Lanczos scalars are inversely proportional to the radial distance. Also since Petrov type D fields have only Coulomb component Ψ_2 of the gravitational field with l^i and n^i as the propagation vectors, Lanczos scalars can act as the potential of the gravitational field, and thus justifying the name—the Lanczos potential. The Vaidya's metric, recently studied by Hasmani and Panchal [21], also exhibit the same feature. A brief account of the work by Hasmani and Panchal is mentioned in the following section.

7.2.5 Vaidya Metric

The Vaidya solution describes the non-empty external spacetime of a spherically symmetric and non-rotating star which is either emitting or absorbing null dusts. It constitutes the simplest non-static generalization of the non-radiative Schwarzschild solution to the Einstein field equations, and therefore also known as 'radiating Schwarzschild metric' or 'the metric of radiating star' (for further details about Vaidya solution, see [1]). The Vaidya metric in null coordinates is given by [1, 46, 47]

$$ds^2 = \left[1 - \frac{2m(u)}{r}\right]du^2 + 2drdu - r^2(d\theta^2 + \sin^2\theta d\phi^2) \qquad (7.23)$$

where $m(u)$ is mass at retarded time u. For this metric, the tetrad vectors are [21]

$$l^i = -\frac{1}{2}\left[1 - \frac{2m(u)}{r}\right]\delta_1^i + \delta_4^i, \ n^i = \delta_1^i, \ m^i = \frac{1}{r\sqrt{2}}(\delta_2^i + i\cos\theta\delta_3^i) \qquad (7.24)$$

and the non-zero spin-coefficients are

$$\rho = \frac{r - 2m(u)}{2r^2}, \ \mu = \frac{1}{r}, \ \epsilon = -\frac{m(u)}{2r^2}, \ \alpha = -\beta = -\frac{\cot\theta}{2\sqrt{2}r} \qquad (7.25)$$

The non-zero components of Weyl scalar and Ricci tensor are, respectively, given by

$$\Psi_2 = -\frac{m(u)}{r^3}, \quad \Phi_{00} = -\frac{m'(u)}{r^2} \tag{7.26}$$

where a dash denotes the partial differentiation with respect to u.

The Lanczos scalars for Vaidya metric are

$$L_0 = L_2 = L_3 = L_4 = L_5 = L_7 = 0, \quad 2L_1 = -L_6 = \frac{1}{3r} \tag{7.27}$$

while using Eq. (5.14), the non-vanishing Lanczos scalars are

$$L_1 = \frac{1}{3}\rho, \quad L_6 = -\frac{1}{3}\mu \tag{7.28}$$

where ρ and μ are given by Eq. (7.25). A comparison of Eqs. (7.27) and (7.28) reveals the non-uniqueness character of Lanczos potential.

7.2.6 Kantowski–Sachs Solution

A solution of Einstein field equations without cosmological constant for dust particles was given by Kantowski and Sachs [24] and the metric in spherical coordinates (r, θ, ϕ, t) is given by

$$ds^2 = dt^2 - X^2(t)dr^2 - Y^2(t)[d\theta^2 + \sin^2\theta d\phi^2] \tag{7.29}$$

The null tetrad vectors for this metric are [22]

$$l^i = \frac{1}{\sqrt{2}}(\delta_4^i + X^{-1}\delta_1^i), \quad n^i = \frac{1}{\sqrt{2}}(\delta_4^i - X^{-1}\delta_1^i), \quad m^i = \frac{1}{Y\sqrt{2}}(\delta_2^i + i\sin^{-1}\theta\delta_3^i) \tag{7.30}$$

The non-vanishing spin-coefficients are given as

$$\rho = -\mu = -\frac{Y'}{Y\sqrt{2}}, \quad \alpha = -\beta = \frac{\cot\theta}{2\sqrt{2}Y}, \quad \epsilon = -\gamma = \frac{X'}{2\sqrt{2}X} \tag{7.31}$$

and the non-zero components of Weyl scalar are

$$\Psi_2 = \frac{1}{3}\left(\frac{Y''}{Y} - \frac{X''}{X}\right)$$

where a dash denotes the differentiation with respect to t.

Therefore, from Eq. (5.14), the Lanczos scalars for Kantowski–Sachs metric are

$$L_0 = L_2 = L_3 = L_4 = L_5 = L_7 = 0, \ L_1 = L_6 = -\frac{Y'}{Y\sqrt{2}} \tag{7.32}$$

which shows that the Lanczos scalars depend only on time as Y is a function of time t.

7.3   Change of Tetrad and Lanczos Potential

Often it is convenient to consider the null tetrad as $Z^i_{(a)} = \{m^i, \bar{m}^i, l^i, n^i\}$ with signature $(+, +, +, -)$ so that the orthogonality conditions are

$$Z^i_{(a)} \ Z_{(b)i} = \begin{bmatrix} 0 & 1 & 0 & 0 \\ 1 & 0 & 0 & 0 \\ 0 & 0 & 0 & -1 \\ 0 & 0 & -1 & 0 \end{bmatrix}$$

From the definition of Ricci rotation coefficients [cf., Eq. (1.15)], now the spin-coefficients are

$$\kappa = -l_{i;j}m^i l^j, \ \tau = -l_{i;j}m^i n^j, \ \sigma = -l_{i;j}m^i m^j, \ \rho = -l_{i;j}m^i \bar{m}^j$$

$$\pi = n_{i;j}\bar{m}^i l^j, \ \nu = n_{i;j}\bar{m}^i n^j, \ \mu = n_{i;j}\bar{m}^i m^j, \ \lambda = n_{i;j}\bar{m}^i \bar{m}^j \tag{7.33a}$$

$$\epsilon = -\frac{1}{2}(l_{i;j}n^i l^j - m_{i;j}\bar{m}^i l^j)$$

$$\gamma = -\frac{1}{2}(l_{i;j}n^i n^j - m_{i;j}\bar{m}^i n^j)$$

$$\beta = -\frac{1}{2}(l_{i;j}n^i m^j - m_{i;j}\bar{m}^i m^j)$$

$$\alpha = -\frac{1}{2}(l_{i;j}n^i \bar{m}^j - m_{i;j}\bar{m}^i \bar{m}^j) \tag{7.33b}$$

Also

$$\delta = m^i \nabla_i, \ \bar{\delta} = \bar{m}^i \nabla_i, \ \Delta = l^i \nabla_i, D = n^i \nabla_i$$

and

$$\nabla_i = l_i \Delta + n_i D - \bar{m}^i \delta - m_i \bar{\delta}$$

The Weyl scalars are

$$\Psi_0 = C_{hijk}n^h m^i n^j m^k$$

$$\Psi_1 = C_{hijk}n^h l^i n^j m^k$$

$$\Psi_2 = -C_{hijk}l^h \bar{m}^i n^j m^k \tag{7.34}$$

$$\Psi_3 = C_{hijk}l^h n^i l^j \bar{m}^k$$

$$\Psi_4 = C_{hijk}l^h \bar{m}^i l^j \bar{m}^k$$

The Lanczos scalars are

$$L_0 = L_{ijk}m^i n^j n^k, \; L_1 = L_{ijk}m^i n^j \bar{m}^k, \; L_2 = L_{ijk}l^i \bar{m}^j n^k, \; L_3 = L_{ijk}l^i \bar{m}^j \bar{m}^k$$

$$L_4 = L_{ijk}m^i n^j m^k, \; L_5 = L_{ijk}m^i n^j l^k, \; L_6 = L_{ijk}l^i \bar{m}^j m^k, \; L_7 = L_{ijk}l^i \bar{m}^j l^k \tag{7.35}$$

The Weyl–Lanczos relations with the choice of tetrad vectors as $\{m^i, \bar{m}^i, l^i, n^i\}$ takes the following form:

$$\Psi_0 = 2[(-\delta + \bar{\alpha} + 3\beta)L_0 + (D - 3\epsilon + \bar{\epsilon})L_4 - \bar{\pi}L_0 - 3\sigma L_1 - \bar{\rho}L_4 + 3\kappa L_5] \tag{7.36a}$$

$$2\Psi_1 = (-\Delta + 3\gamma + \bar{\gamma})L_0 + (-\delta + \bar{\alpha} + \beta)L_1 + (\bar{\delta} - 3\alpha + \bar{\beta})L_4 + (D + \bar{\epsilon} - \epsilon)L_5$$

$$7 + (\mu - \bar{\mu})L_0 - (\tau + \bar{\pi})L_1 - 2\sigma L_2 - (\bar{\tau} + \pi)L_4 + (3\rho - \bar{\rho})L_5 + 2\kappa L_6 \tag{7.36b}$$

$$\Psi_2 = (-\Delta + \gamma + \bar{\gamma})L_1 + (-\delta + \bar{\alpha} - \beta)L_2 + (\bar{\delta} - \alpha + \bar{\beta})L_5 + (D + \bar{\epsilon} + \epsilon)L_6 - \nu L_0$$

$$+ (2\mu - \bar{\mu})L_1 - (2\tau + \bar{\pi})L_2 - \sigma L_3 - \lambda L_4 - (\bar{\tau} + 2\pi)L_5 + (2\rho - \bar{\rho})L_6 + \kappa L_7 \tag{7.36c}$$

$$2\Psi_3 = 3(-\Delta - \gamma + \bar{\gamma})L_2 + (-\delta + \bar{\alpha} - 3\beta)L_3 + 3(\bar{\delta} + \alpha + \bar{\beta})L_6 + (D + \bar{\epsilon} + 3\epsilon)L_7$$

$$+ 6\nu L_1 + 3(\mu - \bar{\mu})L_2 - (3\tau + \bar{\pi})L_3 - 6\lambda L_5 - (3\bar{\tau} + \pi)L_6 + (3\rho - \bar{\rho})L_7 \tag{7.36d}$$

$$\Psi_4 = 2[(-\Delta - 3\gamma + \bar{\gamma})L_3 + (\bar{\delta} + 3\alpha + \bar{\beta})L_7 + 3\nu L_2 - \bar{\mu}L_3 - 3\lambda L_6 - \bar{\tau}L_7] \tag{7.36e}$$

Now once the Lanczos scalars L_i are known, the Lanczos potential can be constructed from the relation

$$L_{ijk} = M_{ijk} + \bar{M}_{ijk} \tag{7.37}$$

where

$$M_{ijk} = L_0 U_{ij}l_k + L_1(W_{ij}l_k - U_{ij}m_k) + L_2(V_{ij}l_k - W_{ij}m_k) - L_3 V_{ij}m_k$$

$$- L_4 U_{ij}\bar{m}_k + L_5(U_{ij}n_k - W_{ij}\bar{m}_k) + L_6(W_{ij}n_k - V_{ij}\bar{m}_k) + L_7 V_{ij}n_k \tag{7.38}$$

with

$$U_{ij} = -l_i\bar{m}_j + l_j\bar{m}_i \ , \quad V_{ij} = n_im_j - n_jm_i$$

$$W_{ij} = -n_il_j + n_jl_i + m_i\bar{m}_j - m_j\bar{m}_i \tag{7.39}$$

Using above considerations, we shall now write down the Lanczos potential for some more well-known solutions of Einstein field equations (see also [24]) and we have

7.3.1 Gödel Solution

Consider the following form of Gödel solution:

$$ds^2 = -(dx^1)^2 - 2e^{x^4}dx^1dx^2 - \frac{1}{2}e^{2x^4}(dx^2)^2 + (dx^3)^2 + (dx^4)^2 \tag{7.40}$$

The null tetrad chosen for this metric is

$$m^i = (1, \ -e^{-x^4}, 0, \ \frac{i}{\sqrt{2}}), \ l^i = \frac{1}{\sqrt{2}}(1, 0, \ -1, 0), \ n^i = \frac{1}{\sqrt{2}}(1, 0, 1, 0) \tag{7.41}$$

so that the non-zero spin-coefficients are

$$\mu = \rho = \frac{i}{2} \tag{7.42}$$

The Lanczos scalars are thus given by [5]

$$L_i = 0, i \neq 1, 6, \quad L_1 = L_6 = \frac{1}{9}\rho \tag{7.43}$$

As Gödel solution is of Petrov type D, so if the scheme (5.14) is used then the Lanczos scalars are

$$L_i = 0, i \neq 1, 6, \quad L_1 = -L_6 = \frac{1}{3}\rho \tag{7.44}$$

It may be noted from Eqs. (7.43) and (7.44) that the Lanczos scalars and consequently Lanczos potential depend upon only on one spin-coefficient either μ or ρ.

7.3.2   Taub Metric

The line element obtained by Taub is [45]

$$ds^2 = \frac{1}{f}[(dx^1)^2 - (dx^4)^2] + f^2[(dx^2)^2 + (dx^3)^2] \tag{7.45}$$

where $f = \sqrt{1 + Kx^1}$, $K = $ constant $\neq 0$.

The appropriate null tetrad vectors for the metric (7.45) are

$$m^i = \frac{1}{\sqrt{2f}}(0,\, 1,\, -i,\, 0),\ l^i = \sqrt{\frac{f}{2}}(-1,\, 0,\, 0,\, 1),\ n^i = \sqrt{\frac{f}{2}}(1,\, 0,\, 0,\, 1) \tag{7.46}$$

The non-zero spin-coefficients are

$$\rho = \mu = 4\epsilon = 4\gamma = -\frac{K}{2\sqrt{2}}f^{-3/2} \tag{7.47}$$

while the Lanczos scalars are [5]

$$L_i = 0, i \neq 1, 6, \quad L_1 = L_6 = \frac{1}{6}\rho \tag{7.48}$$

Moreover, the non-zero component of Weyl scalar is

$$\Psi_2 = \frac{1}{8}K^2 f^{-3} \tag{7.49}$$

which shows that the Taub metric is of Petrov type D and thus using the scheme (5.14), the Lanczos scalars are

$$L_i = 0, \ i = 0, 2, 3, 4, 5, 7, \quad L_1 = -L_6 = -\frac{1}{6\sqrt{2}}Kf^{-3/2} \tag{7.50}$$

It may be noted that if the constant K is taken to be zero, then from Eq. (7.49) $\Psi_2 = 0$, and consequently, Taub spacetime reduces to conformally flat spacetime, and also for $K = 0$ Eq. (7.48)/(7.50) yields $L_i = 0, i = 0, 1, 2, \ldots, 7$.

7.3.3   Petrov Metric

A type D solution obtained by Petrov is described through the metric [41]

$$ds^2 = f^{4/3}[(dx^1)^2 + (dx^2)^2] + f^{-2/3}(dx^3)^2 - (dx^4)^2 \tag{7.51}$$

where $f = ax^4 + 1$, a being a constant.

The null vectors chosen for metric (7.51) are

$$m^i = \frac{1}{\sqrt{2}}(f^{-2/3}, -if^{-2/3}, 0, 0), \; l^i = \frac{1}{\sqrt{2}}(0, 0, -f^{1/3}, 1), \; n^i = \frac{1}{\sqrt{2}}(0, 0, f^{1/3}, 1)$$

$$(7.52)$$

which lead to non-zero spin-coefficients as

$$\rho = -\mu \overset{\angle}{=} 4\epsilon = -\frac{a\sqrt{2}}{3f} \tag{7.53}$$

The Lanczos scalars for the Petrov metric are [5]

$$L_i = 0, i \neq 1, 6, \; \; L_1 = -L_6 = \frac{1}{6}\rho = -\frac{1}{18}\frac{a\sqrt{2}}{f} \tag{7.54}$$

Moreover, the metric (7.51) is of type D as the non-zero component of Weyl scalar is

$$\Psi_2 = -\frac{a^2}{12}f^{-5/3} \tag{7.55}$$

Thus, using the prescription (5.14), we have

$$L_i = 0, \; i = 0, 2, 3, 4, 5, 7, \; \; L_1 = L_6 = -\frac{a\sqrt{2}}{9f} \tag{7.56}$$

It may be noted from Eq. (7.55) that if the constant a is zero then Petrov metric becomes conformally flat and also Eq. (7.54)/(7.56) leads to $L_i = 0$, $i = 0, 1, 2, \ldots, 7$.

7.3.4 Kaigorodov Metric

The metric in (x, y, u, v) coordinates is given by [23]

$$ds^2 = \frac{2}{(ax)^2}(dx^2 + dy^2) - 2du(dv + 2\frac{v}{x}dx + xdu) \tag{7.57}$$

where $a \neq 0$ is a constant. The null vectors for this metric are

$$m^i = a\left(\frac{x}{2}, \frac{ix}{2}, -v, 0\right), \; l^i = \left(0, 0, -\sqrt{x}, \frac{1}{\sqrt{x}}\right), \; n^i = (0, 0, \sqrt{x}, 0) \tag{7.58}$$

and non-zero spin-coefficients are

$$\tau = -\pi = \frac{1}{3}\nu = \frac{1}{2}a \tag{7.59}$$

The Lanczos scalars are

$$L^i = 0, \ i = 0, 1, 2, \ldots, 6, \ \ L_7 = \frac{1}{2}a \tag{7.60}$$

which shows that the Lanczos potential of Kaigorodov metric depends only on the constant a. The non-zero component of Weyl scalar for the metric (7.57) is

$$\Psi_4 = \frac{3}{2}a^2 \tag{7.61}$$

and thus Kaigorodov metric is of Petrov type N. Also, if $a = 0$ then from Eq. (7.61) Kaigorodov metric becomes conformally flat as well as Eq. (7.60) leads to $L_i = 0$ for $i = 0, 1, 2, \ldots, 7$.

Moreover, a comparison of Eqs. (7.60) and (5.57) exhibits the non-uniqueness character of Lanczos potential which has been achieved by different choices of the tetrad vectors.

7.3.5 Kasner Metric

The metric given by Kasner is [25]

$$ds^2 = t^{2p_1}dx^2 + t^{2p_2}dy^2 + t^{2p_3}dz^2 - dt^2 \tag{7.62}$$

where the constants p_i satisfy the conditions

$$p_1 + p_2 + p_3 = 1, \ \ p_1^2 + p_2^2 + p_3^2 = 1$$

The null vectors for the metric (7.62) are

$$m^i = \frac{1}{\sqrt{2}}(t^{-p_1}, -it^{-p_2}, 0, 0), \ l^i = \frac{1}{\sqrt{2}}(0, 0, -t^{-p_3}, 1), n^i = \frac{1}{\sqrt{2}}(0, 0, t^{-p_3}, 1) \tag{7.63}$$

and the only non-zero spin-coefficient is

$$\epsilon = \frac{1}{2\sqrt{2}}\frac{p_3}{t} \tag{7.64}$$

The Lanczos scalars are [5]

$$L_1 = -L_6 = \frac{2}{3}\epsilon, \ \ L_i = 0, \ i \neq 1, 6 \tag{7.65}$$

A comparison of Eqs. (7.65) and (6.51) again shows the non-uniqueness character of Lanczos potential.

7.3.6   Schwarzschild Exterior Solution

Consider the line element

$$ds^2 = A^{-1}dr^2 + r^2(d\theta^2 + \sin^2\theta d\phi^2) - Adt^2 \tag{7.66}$$

where

$$A = 1 - \frac{2M}{r}$$

With the choice of the null vectors

$$m^i = \frac{1}{\sqrt{2r}}(0, 1, -i\csc\theta, 0), \ \ l^i = \frac{1}{\sqrt{2}}(-A^{1/2}, 0, 0, A^{-1/2}),$$

$$n^i = \frac{1}{\sqrt{2}}(A^{1/2}, 0, 0, A^{-1/2}) \tag{7.67}$$

the non-zero spin-coefficients are

$$\gamma = \epsilon = \frac{1}{2\sqrt{2}}Mr^2A^{-1/2} \tag{7.68}$$

The Lanczos scalars for Schwarzschild exterior solution is [24]

$$L_1 = L_6 = \frac{2}{3}\epsilon, \ \ L_i = 0, \ i \neq 1, 6 \tag{7.69}$$

A comparison of Eqs. (7.69) with (7.18) and (7.22) again reveals the non-uniqueness character of Lanczos potential.

7.3.7   C-Metric

This is a vacuum spacetime solution of Einstein field equations which was referred to as C-metric [12]. The static form of this solution was found by Weyl in 1917 [48] and subsequently rediscovered by a number of workers (for further details about C-metric, see [20]). We shall consider the following form of C-metric [in (x, y, ϕ, t) coordinates]

$$ds^2 = (x + y)^{-2}(f^{-1}dx^2 + g^{-1}dy^2 + fd\phi^2 - gdt^2) \qquad (7.70)$$

where $f = x^3 + ax + b$, $g = y^3 + at - b$, a and b are constants.
The null vectors can be taken as

$$m^i = \frac{x + y}{\sqrt{2}}(\sqrt{f}, 0, \frac{i}{\sqrt{f}}, 0), \ l^i = \frac{x + y}{\sqrt{2}}(0, -\sqrt{g}, 0, \frac{1}{\sqrt{g}}),$$

$$n^i = \frac{x + y}{\sqrt{2}}(0, \sqrt{g}, 0, \frac{1}{\sqrt{g}}) \qquad (7.71)$$

so that the non-zero spin-coefficients are

$$\mu = \rho = \sqrt{\frac{g}{2}}, \ \tau = -\pi = \sqrt{\frac{f}{2}}, \ \epsilon = \gamma = -\frac{1}{2}\sqrt{\frac{g}{2}} + \frac{1}{4\sqrt{2g}}(3y^2 + a)(x + y),$$

$$\alpha = -\beta = \frac{1}{2}\sqrt{\frac{f}{2}} + \frac{1}{4\sqrt{2f}}(3x^2 + a)(x + y) \qquad (7.72)$$

The Lanczos scalars for the C-metric are

$$L_0 = -L_7 = -\frac{1}{4}\pi, \ L_1 = L_6 = \frac{1}{3}\epsilon + \frac{1}{12}\rho,$$

$$L_2 = -L_5 = -\frac{1}{3}\beta + \frac{1}{2}\pi, \ L_3 = L_4 = -\frac{1}{4}\rho \qquad (7.73)$$

7.3.8 Siklos Metric

This is a Petrov type III spacetime and the metric in (x, y, r, u) coordinates is [43]

$$ds^2 = r^2 x^{-3}(dx^2 + dy^2) - 2dudr + \frac{3}{2}xdu^2 \qquad (7.74)$$

An appropriate null tetrad for (7.74) is

$$m^i = \frac{x^{3/2}}{r\sqrt{2}}(1, -i, 0, 0), \ l^i = (0, 0, -\frac{\sqrt{3x}}{2}, -\frac{2}{\sqrt{3x}}),$$

$$n^i = (0, 0, -\frac{\sqrt{3x}}{2}, 0) \qquad (7.75)$$

so that the non-zero spin-coefficients are

$$\rho = -\mu = \frac{\sqrt{3x}}{2r}, \quad \beta = -\frac{1}{2}\alpha = -\frac{1}{2}\sqrt{\frac{x}{2r}} \tag{7.76}$$

The Lanczos scalars are

$$L_3 = \frac{1}{4}\mu, \ L_7 = \beta, \ L_i = 0, \ i \neq 3, 7 \tag{7.77}$$

While from Eqs. (5.52) and (7.76), the Lanczos scalars are

$$L_1 = L_6 = \frac{1}{3}\rho, \ L_i = 0, \ i \neq 1, 6 \tag{7.78}$$

which again establishes the non-uniqueness of the Lanczos potential.

Note Using the analysis presented in this section, the Lanczos potential for the Robinson–Trautman metrics of the vacuum and Einstein–Maxwell fields has been obtained by Gaftoi et al. [14].

7.4   Some Other Methods

In the previous chapters, we have discussed the tetrad formalisms for finding the Lanczos potential for a number of spacetime solutions of Einstein field equations. Apart from NP and GHP formalisms, there are other methods too for obtaining the Lanczos potential, and in this section, we shall give three different methods to obtain a Lanczos potential for the Gödel cosmological model [17, 43] with the structure

$$L_{ijk} = Q_{ki;j} - Q_{kj;i}, \ Q_{ij} = Q_{ji} \tag{7.79}$$

which matches with the conditions (4.12) and (4.14) and thus the symmetric tensor Q_{ab} is a generator for the Lanczos tensor L_{ijk} (see also [2, 3]). A similar relation also appears for Kerr geometry [6, 26, 33, 43], plane gravitational waves [34] and Kinnersley metrics [7]. We have the following.

7.4.1   Method of Local and Isometric Embedding

Here we shall show that the Gauss equation employed in the embedding of M_4 into E_5, permits us to find a symmetric tensor b_{ij} which in turn generates a Lanczos potential for the Gödel spacetime.

It is known that [43] a spacetime can be embedded into E_5 if and only if there exists a second fundamental form $b_{ac} = b_{ca}$ satisfying the Gauss–Codazzi equation

$$R_{acij} = \epsilon(b_{ai}b_{cj} - b_{aj}b_{ci}) \tag{7.80}$$

$$B_{cij} = b_{ji;c} - b_{jc;i} \tag{7.81}$$

where $\epsilon = \pm 1$ and R_{acij} is the curvature tensor. In such cases, we say that the space-time is of class one.

From Eq. (7.80), the following identity can easily be obtained [13, 18, 19, 30, 35]

$$pb_{ij} = \frac{R_2}{48}g_{ij} - \frac{1}{2}R_{iacj}S^{ac} \tag{7.82}$$

where

$$S_{ab} = R_{ab} - \frac{1}{2}Rg_{ab}, \quad R_{ab} = R_{afb}{}^f, \quad R = R_a^a$$

are the Einstein tensor, Ricci tensor and scalar curvature, respectively. Also

$$R_2 = {}^*R^{*ijac}R_{ijac}$$

is a Lanczos invariant [16, 28], where

$${}^*R^{*ij}{}_{ac} = \frac{1}{4}\eta^{ijrm}R_{rm}{}^{no}\eta_{noac}$$

η_{ijac} being the Levi-Civita tensor. Moreover

$$p^2 = -\frac{\epsilon}{6}\left(\frac{R}{24}R_2 + R_{imnj}S^{ij}S^{mn}\right) \geq 0 \tag{7.83}$$

If $p \neq 0$ then Eq. (7.82) permits us to obtain explicitly a b_{ij} satisfying Eq. (7.80).
The Gödel metric (signature +2) is given by [17, 43]

$$ds^2 = -(dx^1)^2 - 2e^{x^4}dx^1dx^2 - \frac{1}{2}e^{2x^4}(dx^2)^2 + (dx^3)^2 + (dx^4)^2 \tag{7.84}$$

Applying Eq. (7.82) to Eq. (7.84), we have

$$\epsilon = 1, \quad p = \frac{\sqrt{2}}{4}, \quad (b_{ij}) = -\frac{\sqrt{2}}{2}\begin{bmatrix} 1 & e^{x^4} & 0 & 0 \\ e^{x^4} & \frac{3}{2}e^{2x^4} & 0 & 0 \\ 0 & 0 & 0 & 0 \\ 0 & 0 & 0 & 1 \end{bmatrix} \tag{7.85}$$

The tensor b_{ij} given by Eq. (7.85) does not satisfy Eq. (7.81) as we know that [8, 11, 29, 32, 44] the Gödel metric (7.84) is not of class one, and in fact $b_{12;4} \neq b_{41;2}$. Thus, for this Gödel metric, we have a B_{cij} whose only non-zero independent components are

$$B_{124} = B_{412} = -\frac{\sqrt{2}}{2} e^{x^4}, \; B_{242} = -\frac{3}{2}\sqrt{2}e^{2x^4} \tag{7.86}$$

with the same symmetries as that of Lanczos potential L_{cij} as [10, 16, 39]

$$B_{ijr} = -B_{jir}, \; B_{ijr} + B_{jri} + B_{rij} = 0$$

$$B_i{}^r{}_r = 0 \; \text{(Lanczos algebraic gauge)} \tag{7.87}$$

$$B_{ij}{}^r{}_{;r} = 0 \; \text{(Lanczos differential gauge)}$$

then the following *ansatz* is quite natural

$$L_{ijr} = QB_{ijr}, \; Q = \text{constant} \tag{7.88}$$

which must generate the Weyl tensor through the relation (4.17) with the simplification [33, 40]

$$C_{aijr} = L_{aij;r} - L_{air;j} + L_{jra;i} - L_{jri;a} + g_{ar}L_{ji} - g_{aj}L_{ri} + g_{ij}L_{ra} - g_{ir}L_{ja} \tag{7.89}$$

where $\acute{L}_{ij} = L_i^r{}_{j;r} = L_{ji}'$ because $L_i^r{}_r = 0$.

Using Eqs. (7.86) and (7.88) we find that Eq. (7.89) implies correctly all components of the conformal tensor if $Q = \frac{\sqrt{2}}{18}$, which means that Eq. (7.85) produces a Lanczos potential L_{ijr} for the Gödel spacetime which is given as

$$L_{ijr} = \frac{\sqrt{2}}{18} (b_{rj;i} - b_{ri;j}) \tag{7.90}$$

and is equivalent to Eq. (7.79) with $Q_{ac} = -\frac{\sqrt{2}}{18} b_{ac}$.

It is known that the spacetime (7.84) does not admit embedding into E_5, however, the study of the Gauss–Codazzi equations is important because it allows the construction of the Lanczos generator given by Eq. (7.90) for the Gödel spacetime. Thus, if a metric is not of class one, perhaps, a b_{ij} satisfying Eq. (7.80) may have a relationship similar to Eq. (7.90) with a Lanczos potential for such metric; and the above discussion thus suggests an interesting connection between the embedding of Riemannian 4-space and the Lanczos potential.

7.4.2   Method of the Lovelock's Theorem

The Lanczos algebraic gauge $L_a{}^b{}_b = 0$ may be satisfied if in Eq. (7.79) we set the conditions

$$Q^r{}_r = \text{constant} \tag{7.91}$$

$$Q_a{}^b{}_{;b} = 0 \tag{7.92}$$

Moreover, if we suppose that Q_{ab} depends locally on the intrinsic geometry of M_4, that is

$$Q_{ir} = Q_{ir}(g_{ab}; g_{ab,c}; g_{ab,cd}) \tag{7.93}$$

then the Lovelock's theorem [13, 15, 36, 37] affirms that in four dimensions a tensor of second rank satisfying Eqs. (7.92) and (7.93) must have the form

$$Q_{ab} = \alpha S_{ab} + \beta g_{ab} \tag{7.94}$$

where α and β are constants. Thus, $Q_r{}^r = 4\beta - \alpha R$ satisfies Eq. (7.91) because for the Gödel metric (7.84), $R = 1$; and the substitution of Eq. (7.94) into the Eq. (7.79) leads to

$$L_{ijk} = \alpha(R_{ki;j} - R_{kj;i}) \tag{7.95}$$

Finally, from Eqs. (7.84), (7.89) and (7.95), we conclude that $\alpha = \frac{1}{9}$, and thus Eq. (7.95) becomes

$$L_{ijk} = \frac{1}{9}(R_{kj;i} - R_{ki;j}) \tag{7.96}$$

which is same as Eq. (7.90) because b_{ac} is connected with R_{ac} through the relation

$$b_{ij} = \sqrt{2}\left[R_{ij} + \frac{1}{2}(B_i B_j - g_{ij})\right] \tag{7.97}$$

where B_j is a constant space-like vector given by

$$(B_j) = (0, 0, 1, 0), \quad B_{j;i} = 0 \tag{7.98}$$

It is interesting to note that

$$L_{ijk} v^k \propto \omega_{ij}$$

where $(v^i) = (1, 0, 0, 0)$ is the velocity of the fluid and ω_{ij} is the spin [42] of the matter rotating in this Gödel spacetime, which is in accordance with the proposition of [31] that L_{ijr} represents some type of angular momentum into the spacetime.

7.4.3 Method of the Wave Equation

Here, we shall consider a second rank symmetric tensor satisfying the wave equation (cf., [3])

$$W_{ab}{}^{;r}{}_{;r} = 0 \tag{7.99}$$

in the Gödel spacetime, and find a solution of the wave equation (7.99) which generates a Lanczos potential for Gödel spacetime with the structure given by Eq. (7.79).

In fact, Eq. (7.99) admits the solution

$$(W_{ab}) = \frac{1}{27} \begin{bmatrix} 4 & 4e^{x^4} & 0 & 0 \\ 4e^{x^4} & \frac{7}{2}e^{2x^4} & 0 & 0 \\ 0 & 0 & 0 & 0 \\ 0 & 0 & 0 & -1 \end{bmatrix}, \quad W_a{}^b{}_{;b} = 0 \tag{7.100}$$

then

$$L_{ij} = W_{ki;j} - W_{kj;i} \tag{7.101}$$

implies a Lanczos generator whose non-zero independent components are

$$L_{124} = L_{412} = -\frac{e^{x^4}}{18}, \quad L_{242} = \frac{e^{2x^4}}{6} \tag{7.102}$$

with all properties described by Eq. (7.87), which reproduces, via Eq. (7.89), correctly the Weyl tensor. The Lanczos potential (7.101) is equivalent to Eq. (7.90) due to the relation

$$W_{ri} = \frac{1}{18} \left[-\sqrt{2}b_{ri} + \frac{5}{3}(B_r B_i - g_{ri}) \right] \tag{7.103}$$

Thus, the above discussion clearly shows that it is important to study the wave Eq. (7.99) for the construction of a Lanczos potential in Gödel geometry. However, it may be noted that this situation may appear in other spacetimes too.

References

1. Ahsan, Z.: Tensors: Mathematics of Differential Geometry and Relativity. Second Printing, PHI Learning Pvt. Ltd, Delhi (2017)
2. Ahsan, Z., Caltenco, J.H., Lopez-Bonilla, J.L.: Annalen der Physik 16(4), 311–313 (2007)
3. Ahsan, Z., Caltencho, J.H., Linares Y.M.R., Lopez-Bonilla, J.: Commun. Phys. 20(1), 9–14 (2010)
4. Ali, M., Ahsan, Z.: SUT J. Maths. 49(2), 129–143 (2013)
5. Ares de Parga, G., Lopez Bonilla, J., Ovando, G., Matos, T.: Rev. Mex. Fis. 35(3), 393–409 (1989)
6. Caltenco, J.H., Lopez-Bonilla, J., Morales, J., Ovando, G.: Chinese. J. Phys. 39, 397–400 (2001)
7. Caltenco, J.H., Lopez-Bonilla, J., Ovando, G., Rivera, J.: Apeiron 9, 38–42 (2002)
8. Collinson, C.D.: J. Math. Phys. 9, 403–410 (1968)
9. Dolan, P., Kim, C.W.: Proc. Roy. Soc. Lond. A 447, 577–585 (1994)
10. Edgar, S.B., Höglund, A.: Proc. Roy. Soc. Lond. A 453, 835–851 (1997)
11. Eguchi, K.: J. Math. Tokoshima Univ. 1, 17–27 (1967)
12. Ehler, J., Kundt, W.: Article in Gravitation: An Introduction to Current Research. Edited by Luis Witten (1962)

13. Fuentes, R., Lopez-Bonilla, J., Ovando, G., Matos, T.: Gen. Rel. Grav. **21**, 777–784 (1989)
14. Gaftoi, V., Morales, J., Ovando, G., Peña, J.J.: Nuovo Cimento B **113**, 1297 (1998)
15. Gaftoi, V., Lopez-Bonilla, J., Ovando, G.: Int. J. Theo. Phys. **38**, 939–943 (1999)
16. Gaftoi, V., Lopez-Bonilla, J., Ovando, G.: Nuovo Cimento B **113**, 1489–1492 (1998)
17. Gödel, K.: Rev. Mod. Phys. **21**, 447–450 (1949)
18. Goenner, H.F.: Tensor (N.S.) **30**, 15 (1976)
19. Gonzalez, G., Lopez-Bonilla, J., Rosales, M.: Pramana. J. Phys. **42**, 85–88 (1994)
20. Griffiths, J.B., Krtous, P., Podolsky, J.: Interpreting the C-metric. arXiv:gr-qc/0609065v1 (15 Sept. 2006)
21. Hasmani, A.H., Panchal, R.: Eur. Phys. J. Plus **131**, 336–341 (2016)
22. Ibohal, N.G.: Astrophys. Space Sci. **249**, 73–93 (1997)
23. Kaigorodov, V.R.: Sov. Phys. Doklady **7**, 893 (1963)
24. Kantowski, R., Sachs, R.K.: J. Math. Phys. **7**, 443 (1966)
25. Kasner, E.: Am. J. Math. **43**, 217 (1921)
26. Kerr, R.P.: Phys. Rev. Lett. **11**, 237–238 (1963)
27. Kinnersley, W.: J. Math. Phys. **10**, 1195 (1969)
28. Lanczos, C.: Ann. Math. **39**, 842–850 (1938)
29. Lopez-Bonilla, J., Morales, J., Rosales, M.: Braz. J. Phys. **24**, 522–525 (1994)
30. Lopez-Bonilla, J., Nunez-Yepez, H.N.: Pramana. J. Phys. **46**, 219–221 (1996)
31. Lopez-Bonilla, J., Ovando, G., Rivera, J.: Nuovo Cimento B **112**, 1433–1436 (1997)
32. Lopez-Bonilla, J., Ovando, G., Rivera, J.: Alig. Bull. Maths. **17**, 63–66 (1997–98)
33. Lopez-Bonilla, J., Morales, J., Ovando, G.: Gen. Rel. Grav. **31**, 413–415 (1999)
34. Lopez-Bonilla, J., Ovando, G., Pena, J.J.: Found. Phys. Lett. **12**, 401–405 (1999)
35. Lopez-Bonilla, J., Morales, J., Ovando, G.: Indian J. Maths. **42**, 309–312 (2000)
36. Lopez-Bonilla, J., Morales, J., Ovando, G.: Indian. J. Phys. B **74**, 397–398 (2000)
37. Lovelock, D.: J. Math. Phys. **13**, 874–876 (1972)
38. Newman, E.T., Couch, E., Chinnapared, K., Exton, E., Prakash, A., Torrence, R.: J. Math. Phys. **6**, 918–919 (1965)
39. O'Donell, P.J.: Gen. Rel. Grav. **36**, 1415–1422 (2004)
40. O'Donell, P.J.: Czech. J. Phys. **54**, 889–896 (2004)
41. Petrov, A.Z.: Recent Developments in General Relativity. Pergamon Press (1962)
42. Raychaudhuri, A.: Phys. Rev. **98**, 1123–1126 (1955)
43. Stephani, H., Kramer, D., Maccallum, M., Hoenselaers, C., Herlt, E.: Exact Solutions of Einstein's Field Equations, 2nd edn. Cambridge University Press, UK (2003)
44. Szekeres, P.: Nuovo Cimento **A43**, 1062 (1966)
45. Taub, A.H.: Ann. Math. **53**, 472 (1951)
46. Vaidya, P.C.: Curr. Sci. **12**, 12 (1943)
47. Vaidya, P.C.: Proc. Indian Acad. Sci. **33**, 264 (1951)
48. Weyl, H.: Ann. Physik **54**, 117 (1917)
49. Zund, J.D.: Ann. Mat. Pure Appl. **104**, 239–268 (1975)

Chapter 8
Newman–Penrose Formalism, Solution of Einstein–Maxwell Equations and Symmetries of the Spacetime

8.1 Introduction

In Chap. 2, we have presented a detailed account of the Newman–Penrose formalism and have applied this formalism to study Lanczos potential in subsequent chapters. In this chapter, we shall deal with some other applications of NP formalism. In Sect. 8.2, we shall consider the interaction of a Petrov type N gravitational field with a null electromagnetic field and obtain a metric describing such a situation. The geometric and physical properties of the solution have also been discussed. While in Sect. 8.3, we shall be concerned with the geometrical symmetries corresponding to the continuous groups of motions generated by a null vector. For Petrov type N pure radiation fields, these symmetries have been studied in detail by using NP formalism.

8.2 A Solution of Einstein–Maxwell Equations

During the last four decades, it is seen that the Petrov type N solutions of Einstein vacuum field equations are among the most interesting, but rather difficult and little explored of all empty space metrics [29]. From the physical point of view, they represent spacetime filled up entirely with gravitational radiation, while mathematically they form a class of solutions of Einstein equations which should be possible to be determined explicitly. The behaviour of gravitational radiation from a bounded source is an important problem. Moreover, due to the detection of gravitational waves in recent times, the study of type N solutions becomes more pertinent.

Here we shall consider the free gravitational field to be the transverse gravitational wave zone which can be identified as Petrov type N fields [30] and focus our attention on the interaction of pure electromagnetic radiation field and pure gravitational radiation field. We call such fields as pure radiation fields (or PR fields). For this study, we shall be using the terminology of Chap. 2 and consider the vector l^i of the

© Springer Nature Singapore Pte Ltd. 2019
Z. Ahsan, *The Potential of Fields in Einstein's Theory of Gravitation*,
https://doi.org/10.1007/978-981-13-8976-4_8

null tetrad $\{l^i, n^i, m^i, \bar{m}^i\}$ as the common propagation vector of the PR fields [cf., Eqs. (8.55), (8.65) and (8.72)].

It is known that a vacuum metric contains a geodesic ray if and only if there is a principal null direction of the curvature tensor such that

$$l_{[p} R_{q]rs[t} l_{u]} l^r l^s = 0 \tag{8.1}$$

i.e. tangent to a congruence of null geodesics

$$l_{i;j} l^j = 0, \quad l_i l^i = 0 \tag{8.2}$$

Assume that the geodesic rays are hypersurface orthogonal, i.e. $l_i = u_{,i}$. Introduce the coordinates $u = x^1, r = x^2$ and x^3, x^4, where r is an affine parameter (a parameter along the geodesic such that the equation of the geodesic takes the standard form) along the null geodesic and x^3, x^4 label the geodesic on each surface $u = $ constant. The metric has the form

$$g^{ij} = \begin{bmatrix} 0 & 1 & 0 & 0 \\ 1 & g^{22} & g^{23} & g^{24} \\ 0 & g^{23} & g^{33} & g^{34} \\ 0 & g^{24} & g^{34} & g^{44} \end{bmatrix}. \tag{8.3}$$

The null vectors have the form

$$l^i = \delta_2^i, \quad n^i = \delta_1^i + U\delta_2^i + X^i\delta_1^i, \quad m^i = \omega\delta_2^i + \xi^j\delta_j^i \tag{8.4}$$

The components of the metric tensor are

$$g^{22} = 2(U - \omega\bar{\omega}), \quad g^{2i} = X^i - (\xi^i\bar{\omega} + \bar{\xi}^i\omega), \quad g^{ij} = -(\xi^i\bar{\xi}^j + \bar{\xi}^i\xi^j), \quad i, j = 3, 4 \tag{8.5}$$

and the intrinsic derivative operators are given by

$$D = \frac{\partial}{\partial r}, \quad \Delta = U\frac{\partial}{\partial r} + \frac{\partial}{\partial u} + X^i\frac{\partial}{\partial x^i}, \quad \delta = \omega\frac{\partial}{\partial r} + \xi^i\frac{\partial}{\partial x^i} \tag{8.6}$$

It can be seen that the metric tensor g^{ij} remain invariant under the coordinate transformation

$$r' = r + R(1, 3, 4), \quad u' = u, \quad x^{i'} = x^i \quad \text{(shifts the origin)} \tag{8.7a}$$

$$r' = r/\dot{r}, \quad u' = r(u), \quad x^{i'} = x^i \quad \text{(relables hypersurface)} \tag{8.7b}$$

$$r' = r, \quad u' = u, \quad x^{i'} = x^i \quad \text{(relables hypersurface)} \tag{8.7c}$$

The orthogonality properties of the tetrad vectors remain unchanged under the transformation

$$l^{i'} = l^i, \ n^{i'} = n^i + \bar{P}m^i + P\bar{m}^i + P\bar{P}l^i, \ m^{i'} = m^i + Pl^i \tag{8.8a}$$

and

$$l^{i'} = l^i, \ n^{i'} = n^i, \ m^{i'} = m^i e^{iQ} \tag{8.8b}$$

where P is a complex scalar and Q is a real independent of r. Following coordinate freedom shall also be used:

$$r' = r/\dot{r}, \ u' = r(u), \ \zeta' = x^{3'} + ix^{4'} = f(\zeta, u) \tag{8.9}$$

where $\zeta = x^3 + ix^4$.

From Chap. 2, Petrov type N gravitational field, having l^i as the propagation vector, is characterized by the condition

$$\Psi_4 = \Psi \neq 0, \ \Psi_i = 0, \ i = 0, 1, 2, 3 \tag{8.10}$$

while Maxwell scalar which characterizes the null (radiative) electromagnetic field having l^i as the propagation vector is given by [16]

$$\Phi_2 = \Phi \neq 0, \ \Phi_i = 0, \ i = 0, 1 \tag{8.11}$$

and the electromagnetic field tensor is

$$F_{ij} = \Phi l_{[i}m_{j]} + \bar{\Phi}l_{[i}\bar{m}_{j]} \tag{8.11a}$$

where $\Phi = 2F_{ij}\bar{m}^i n^j$, a square bracket denotes the skew-symmetrization and a parenthesis denotes the symmetrization.

As we working with PR fields, therefore, from Eq. (8.10) and the Goldberg–Sachs theorem, the propagation vector l^i is geodetic and shear-free; moreover from Eq. (8.11) we see that the vector l^i is scaled. We thus have

$$\kappa = \sigma = \epsilon = 0 \tag{8.12}$$

8.2.1 Equations for PR Fields and Their Simplifications

The field Eqs. (2.33), the commutator relations (2.28–2.31), the Bianchi identities (2.34), Maxwell equations [16, 25] the coupled Bianchi identities [29], using Eqs. (8.10)–(8.12), reduce to the following form [5]:

$$D\rho = \rho^2 \tag{8.13a}$$

$$D\tau = (\tau + \bar{\pi})\rho \tag{8.13b}$$

$$D\alpha = (\alpha + \pi)\rho \tag{8.13c}$$

$$D\beta = \beta\bar{\rho} \tag{8.13d}$$

$$D\gamma = (\tau + \bar{\pi})\alpha + (\bar{\tau} + \pi)\beta + \tau\pi \tag{8.13e}$$

$$D\lambda - \bar{\delta}\pi = \rho\lambda + \pi^2 + (\alpha - \bar{\beta})\pi \tag{8.13f}$$

$$D\mu - \delta\pi = \bar{\rho}\mu + \pi\bar{\pi} - \pi(\bar{\alpha} - \beta) \tag{8.13g}$$

$$D\nu - \Delta\pi = (\pi + \bar{\tau})\mu + (\bar{\pi} + \tau)\lambda + (\gamma - \bar{\gamma})\pi \tag{8.13h}$$

$$\Delta\lambda - \bar{\delta}\nu = -(\mu + \bar{\mu})\lambda - (3\gamma - \bar{\gamma})\lambda + (3\alpha + \bar{\beta} + \pi - \tau)\nu - \Psi \tag{8.13i}$$

$$\delta\rho = \rho(\bar{\alpha} + \beta) + (\rho - \bar{\rho})\tau \tag{8.13j}$$

$$\delta\alpha - \bar{\delta}\beta = \rho\mu + \alpha\bar{\alpha} + \beta\bar{\beta} - 2\alpha\beta + \gamma(\rho - \bar{\rho}) \tag{8.13k}$$

$$\delta\lambda - \bar{\delta}\mu = \pi(\mu - \bar{\mu}) + \mu(\alpha + \bar{\beta}) + \lambda(\bar{\alpha} - 3\beta) - \nu(\rho - \bar{\rho}) \tag{8.13l}$$

$$\delta\nu - \Delta\mu = \mu^2 + \lambda\bar{\lambda} + \mu(\gamma + \bar{\gamma}) - \bar{\nu}\pi + \nu(\tau - 3\beta - \bar{\alpha}) + \Phi\bar{\Phi} \tag{8.13m}$$

$$\delta\gamma - \Delta\beta = \gamma(\tau - \bar{\alpha} - \beta) + \mu\tau - \beta(\gamma - \bar{\gamma} - \mu) + \alpha\bar{\lambda} \tag{8.13n}$$

$$\delta\tau = \rho\bar{\lambda} + \tau(\tau + \beta - \bar{\alpha}) \tag{8.13o}$$

$$\Delta\rho - \bar{\delta}\tau = -\rho\bar{\mu} + \tau(\bar{\beta} - \alpha - \bar{\tau}) + \rho(\gamma + \bar{\gamma}) \tag{8.13p}$$

$$\Delta\alpha - \bar{\delta}\gamma = \rho\nu - \lambda(\tau + \beta) - \alpha(\bar{\gamma} - \mu) + \gamma(\bar{\beta} - \bar{\tau}) \tag{8.13q}$$

$$(\Delta D - D\Delta)\eta = [(\gamma + \bar{\gamma})D - (\tau + \bar{\pi})\bar{\delta} - (\bar{\tau} + \pi)\delta]\eta \tag{8.14a}$$

$$(\delta D - D\delta)\eta = [(\bar{\alpha} + \beta - \pi)D - \bar{\rho}\delta]\eta \tag{8.14b}$$

$$(\delta\Delta - \Delta\delta)\eta = [-\bar{\nu}D + (\tau - \bar{\alpha} - \beta)\Delta + \lambda\bar{\delta} + (\mu - \gamma + \bar{\gamma})\delta]\eta \tag{8.14c}$$

$$(\bar{\delta}\delta - \delta\bar{\delta})\eta = [(\bar{\mu} - \mu)D - (\bar{\alpha} - \beta)\bar{\delta} - (\bar{\beta} - \alpha)\delta]\eta \tag{8.14d}$$

$$D\Psi = \rho\Psi \tag{8.15a}$$

$$\delta\Psi = -4\beta\Psi \tag{8.15b}$$

$$D\phi = \rho\Phi \tag{8.16a}$$

$$\delta\Phi = (\tau - 2\beta)\Phi \tag{8.16b}$$

$$D(\Phi\bar{\Phi}) = -\rho\Phi\bar{\Phi} \tag{8.17a}$$

$$\delta(\Phi\bar{\Phi}) - \delta\Psi = (-\tau + 4\beta)\Psi - 2(\bar{\beta} + \alpha)\Phi\bar{\Phi} \tag{8.17b}$$

From Eqs. (8.6) and (8.14) we have [replacing η by u, r, x^3, x^4 on each equation of (8.14)]

$$DU = -(\gamma + \bar{\gamma}) + (\tau + \bar{\pi})\bar{\omega} + (\bar{\tau} + \pi)\omega \tag{8.18a}$$

$$DX^i = (\tau + \bar{\pi})\bar{\xi}^i + (\bar{\tau} + \pi)\xi^i \tag{8.18b}$$

$$D\omega = -(\bar{\alpha} + \beta - \bar{\pi}) + \bar{\rho}\omega \tag{8.18c}$$

$$D\xi^i = \bar{\rho}\xi^i \tag{8.18d}$$

$$\tau = \bar{\alpha} + \beta \quad [\text{taking } \eta = u \text{ in Eq. (8.14c)}] \tag{8.18e}$$

$$\delta U - \Delta\omega = \bar{\nu} + (\tau - \bar{\alpha} - \beta)U + \bar{\lambda}\bar{\omega} + (\mu - \gamma + \bar{\gamma})\omega \tag{8.18f}$$

$$\delta X^i - \Delta\xi^i = \lambda\bar{\xi}^i + (\bar{\tau} - \bar{\alpha} - \beta)X^i + (\mu - \gamma + \bar{\gamma})\xi^i \tag{8.18g}$$

$$\rho = \bar{\rho} \quad [\text{taking } \eta = u \text{ in Eq. (8.14d)}] \tag{8.18h}$$

$$\bar{\delta}\omega - \delta\bar{\omega} = (\bar{\mu} - \mu) - (\bar{\alpha} - \beta)\bar{\omega} - (\bar{\beta} - \alpha)\omega \tag{8.18i}$$

$$\bar{\delta}\xi^i - \delta\bar{\xi}^i = -(\bar{\alpha} - \beta)\bar{\xi}^i - (\bar{\beta} - \alpha)\xi^i \tag{8.18j}$$

In our coordinate system, from Eq. (8.18e), $\tau^o = \bar{\alpha}^o + \beta^o$, $\tau^o = 0$ by Eq. (8.8) and thus

$$\alpha = -\bar{\beta}, \ \tau = 0 \tag{8.19}$$

From Eqs. (8.18h) and (8.19), Eqs. (8.13b), (8.13j) and (8.13o) lead to

$$\pi = 0 \tag{8.20a}$$

$$\delta\rho = 0 \implies \omega\frac{\partial\rho}{\partial r} = \omega D\rho = 0 \implies \omega = 0 \tag{8.20b}$$

and

$$\lambda = 0 \tag{8.20c}$$

From Eqs. (8.18h), (8.19) and (8.20), after simplifications, the radial (equations depending on r) and non-radial equations are given as follows.

Radial Equations

$$D\rho = \rho^2 \tag{8.21a}$$

$$D\alpha = \rho\alpha \tag{8.21b}$$

$$D\beta = \rho\beta \tag{8.21c}$$

$$D\gamma = 0 \tag{8.21d}$$

$$D\mu = \rho\mu \tag{8.21e}$$

$$D\nu = 0 \tag{8.21f}$$

$$D\Psi = \rho\Psi \tag{8.21g}$$

$$D\Phi = \rho\Phi \tag{8.21h}$$

$$DU = -(\gamma + \bar{\gamma}) \tag{8.21i}$$

$$DX^i = 0 \tag{8.21j}$$

$$D\xi^i = \rho\xi^i \tag{8.21k}$$

Non-Radial Equations

$$\bar{\delta}\nu = -2\alpha\nu + \Psi \tag{8.22a}$$

$$\delta\rho = 0 \tag{8.22b}$$

$$\delta\alpha + \bar{\delta}\bar{\alpha} = \rho\mu + 4\alpha\bar{\alpha} \tag{8.22c}$$

$$\bar{\delta}\mu = 0 \tag{8.22d}$$

$$\delta\nu - \Delta\mu = \mu^2 + \mu(\gamma + \bar{\gamma}) + 2\bar{\alpha}\nu + \Phi\bar{\Phi} \tag{8.22e}$$

$$\delta\gamma + \Delta\bar{\alpha} = \bar{\mu}(\gamma - \bar{\gamma} - \mu) \tag{8.22f}$$

$$\Delta\rho = \rho(\gamma + \bar{\gamma} - \bar{\mu}) \tag{8.22g}$$

$$\Delta\alpha - \bar{\delta}\gamma = \rho\nu + \alpha(\bar{\gamma} - \gamma - \mu) \tag{8.22h}$$

$$\delta\Psi = 4\bar{\alpha}\Psi \tag{8.22i}$$

$$\delta\Phi = 2\bar{\alpha}\Phi \tag{8.22j}$$

$$\delta U = -\bar{\nu} \tag{8.22k}$$

$$\delta X^i - \Delta\xi^i = (\mu - \gamma + \bar{\gamma})\xi^i \tag{8.22l}$$

$$\bar{\delta}\xi^i - \delta\bar{\xi}^i = -2\bar{\alpha}\bar{\xi}^i - 2\alpha\xi^i \tag{8.22m}$$

$$\bar{\delta}(\Phi\bar{\Phi}) - \delta\Psi = -4\bar{\alpha}\Psi \tag{8.22n}$$

$$\bar{\delta}(\Phi\bar{\Phi}) = 0 \text{ [using Eq. (8.22i) in Eq. (8.22n)]} \tag{8.22o}$$

The solution of radial Eq. (8.21a) is $\rho = (-r + \rho^o)^{-1}$ where ρ^o is the constant of integration and is equal to zero from Eq. (8.7a). A degree sign as upper index indicates that the function is independent of r and thus

$$\rho = -\frac{1}{r} \tag{8.23a}$$

The solutions of the remaining radial equations are as follows:

$$\alpha = \frac{\alpha^o}{r} \tag{8.23b}$$

$$\beta = \frac{\beta^o}{r} \tag{8.23c}$$

$$\gamma = \gamma^o \tag{8.23d}$$

$$\mu = \frac{\mu^o}{r} \tag{8.23e}$$

$$\nu = \nu^o \tag{8.23f}$$

$$\Psi = \frac{\Psi^o}{r} \tag{8.23g}$$

$$\Phi = \frac{\Phi^o}{r} \tag{8.23h}$$

$$U = U^o - (\gamma^o + \bar{\gamma}^o)r \tag{8.23i}$$

$$X^i = X^{oi} \tag{8.23j}$$

$$\xi^i = \frac{\xi^{oi}}{r} \tag{8.23k}$$

From Eq. (8.7) we can write $\xi^{oj} = a(\delta_3^j + \delta_4^j)$ and the function $a \equiv a(u, \zeta, \bar{\zeta})$ is real. Using the above results into the non-radial Eq. (8.22l) and comparing the coefficients of like powers of $1/r$, we get

$$\mu^o = U^o \tag{8.24}$$

$$\xi^{oj}(X_{,j}^{o3} + i X_{,j}^{o4}) \equiv a\nabla X^o = 0 \tag{8.25}$$

where $\nabla \equiv \dfrac{\partial}{\partial x^3} + i \dfrac{\partial}{\partial x^4}$, $X^o = X^{o3} + i X^{o4}$, and from Eq. (8.9), we have

$$X^{oi} = 0 \tag{8.26}$$

Moreover, Eqs. (8.26) and (8.22l) lead to

$$2\gamma^o = \frac{\partial}{\partial u} \log a \tag{8.27}$$

In a similar manner, the other non-radial equations can be worked out [as that of Eq. (8.22l)] and we have

$$\text{Eq. (8.22k)} \implies \bar{\nu}^o = \frac{1}{2} a\nabla\left(\frac{\partial}{\partial u} \log a\bar{a}\right) \tag{8.28}$$

$$\text{Eq. (8.22m)} \implies 2a^o = \bar{a}\bar{\nabla} \log a \tag{8.29}$$

$$\text{Eq. (8.22c)} \implies \mu^o = U^o = 4\alpha^o\bar{\alpha}^o = a\bar{a}\nabla\bar{\nabla} \log a \log \bar{a} \tag{8.30}$$

$$\text{Eq. (8.22a)} \implies \Psi^o = \bar{a}\nabla\nu^o + 2\alpha^o\nu^o \tag{8.31}$$

$$\text{Eq. (8.22i)} \implies a\nabla\Psi^o = 4\bar{\alpha}^o\Psi^o \tag{8.32}$$

$$\text{Eq. (8.22j)} \implies a\nabla\Phi^o = 2\alpha^o\Phi^o \tag{8.33}$$

while Eq. (8.22o) leads to

$$\bar{a}\bar{\nabla}\Phi^o\bar{\Phi}^o = 0$$

which yield

$$\nabla\Phi^o\bar{\Phi}^o = \bar{\nabla}\Phi^o\bar{\Phi}^o = 0$$

and thus

$$\Phi^o = \Phi^o(u) \tag{8.34}$$

Using Eq. (8.8) we can set $a = \bar{a}$ and thus

$$\alpha^o = \frac{1}{2}aA \tag{8.35}$$

$$\mu^o = U^o = a^2 B \tag{8.36}$$

$$\gamma^o = \frac{1}{2}C \tag{8.37}$$

$$\nu^o = aE \tag{8.38}$$

$$\Psi^o = a^2 \nabla (1 + \log a) E \tag{8.39}$$

where

$$A = \nabla \log a, \ \ B = \nabla^2 (\log a)^2, \ \ C = \frac{\partial}{\partial u} \log a \ \text{ and } \ E = \nabla C = \nabla \frac{\partial}{\partial u} \log a$$

The remaining non-radial are identically satisfied and hence the integration of all the quantities is completed.

8.2.2   The Solution Describing the Interaction of PR Fields

Using Eq. (8.5) and the simplifications given in the above section, the spacetime representing the interaction of pure radiation fields is described through the metric

$$ds^2 = Fdu + 2dudr + g_{ij}dx^i dx^j \tag{8.40}$$

where $F = 2[a^2 B - Cr]$ and $g^{ij} = (g_{ij})^{-1} = -\frac{2a^2}{r^2} \delta^{ij}$. Here the propagation vector l^i of the two interacting fields is geodetic, shear-free, non-expanding and hypersurface orthogonal and the tetrad $\{l^i, n^i, m^i, \bar{m}^i\}$ is parallelly propagated along l^i. The interaction of the gravitational radiation with electromagnetic radiation is described by the Eq. (8.22n). The solution (8.40) also has the singularities at $r = 0$ and $r \to \infty$.

It may be noted that the shear-free (i.e. $\sigma = 0$) and geodetic (i.e. $\kappa = 0$) properties of the propagation vector l^i are due to the interaction of PR fields., while the parallel propagation of the tetrad (cf., Theorem 2.8) and the twist-free (i.e. $\rho - \bar{\rho} = 0$) properties of the propagation vector have been achieved as a consequence of field equations, commutation relations and their simplification as described in Sect. 8.2.1.

The solution (8.40) reduces to the spacetime solution representing the spherical gravitational waves and the Siklos metric [28] if $\gamma = \nu = 0$, $\rho \neq \bar{\rho}$ and there is no electromagnetic radiation; while the spacetime solution (8.40) becomes the well-known Kerr twisting spacetime [31] if $\rho \neq \bar{\rho}$, $\Psi_i \neq 0$, $i = 2, 3, 4$. However, the choice of the tetrad vectors (8.4) and accordingly the coordinate transformation will be different from those discussed here.

8.3   Symmetries of Type N Pure Radiation Fields

In general theory of relativity, the curvature tensor describing the gravitational field consists of two parts, viz. the matter part and the free gravitational part. The interaction between these two parts is described through Bianchi identities. For a given distribution of matter, the construction of gravitational potential satisfying Einstein's field equations is the principal aim of all investigations in gravitational physics and this has often been achieved by imposing symmetries on the geometry compatible with the dynamics of the chosen distribution of matter. These geometrical symmetries of the spacetime are often defined by the vanishing of the Lie derivative of certain tensors with respect to a vector (this vector may be time-like, space-like or null).

The geometrical symmetries of the spacetime are expressible through the equation

$$\pounds_\xi A = 2\Omega A \tag{8.41}$$

where A represents a geometrical/physical quantity, $\pounds_\xi$ denotes the Lie derivative of A with respect to the vector field ξ and Ω is a scalar. One of the most simple and widely used examples is the metric inheritance symmetry for which $A = g_{ij}$ in Eq. (8.41). In this case, the vector field ξ^a is called the conformal Killing vector which includes a homothetic vector and a Killing vector according as $\Omega_{;a} = 0$ and $\Omega = 0$, respectively.

In recent years, there has been much interest in symmetries of spacetime in general relativity. Such interest is due to the need to simplify the Einstein's field equations in the search for their exact solutions. In a series of papers [14, 15, 20–22] Katzin, Levine, Davis and collaborators have identified a number of symmetries for the gravitational field with their interrelationships and obtained the corresponding weak conservation laws as the integrals of the geodesic equation. Different types of matter distribution compatible with geometrical symmetries have been the subject of interest of several investigators for quite some time and in this connection, Oliver and Davis [27], for the perfect fluid spacetimes, have studied the time-like symmetries with special reference to conformal motion and family of contracted Ricci collineation. The perfect fluid spacetimes including electromagnetic field which admit symmetry mapping belonging to the family of contracted Ricci collineation, have been studied by Norris et al. [26]. The role of geometrical symmetries in the study of fluid spacetimes, with an emphasis on conformal collineation, has been explored by Duggal [18] and Duggal and Sharma [19] (see also [4]). The geometrical symmetry $\pounds_\xi R_{ij} = 2\Omega R_{ij}$, known as Ricci inheritance, has been studied by Ahsan [6] who obtained the necessary and sufficient conditions for perfect fluid spacetimes to admit such symmetries in terms of the kinematical quantities. These symmetries which are also known as collineations were further studied by Ahsan [1–6], Ahsan and Ahsan [7], Ahsan and Hussain [10], Ahsan and Ali [8, 9] and Ali and Ahsan [11, 12] among many others.

Different types of symmetries of Petrov type N gravitational fields have been the subject of interest since last few decades (cf., [5]) but a complete analysis of collineations is not found in the literature (as far as we know). Here, we consider the free gravitational field to be the transverse gravitational wave zone which can be identified as Petrov type N fields and focus our attention on the interaction of pure electromagnetic radiation field and pure gravitational radiation field. For such PR fields, using NP formalism [25], a systematic and detailed study of different types of collineations has been made.

8.3.1 NP Formalism and Collineations for PR Fields

Consider a four-dimensional Lorentzian manifold M which admits a Lorentzian metric g_{ij}. Let $Z_\mu^a = \{l^a, n^a, m^a, \bar{m}^a\}$ be the complex null tetrad ($\mu = 1, 2, 3, 4$) where l^a, n^a are real null vectors and m^a, $\bar{m}^a$ are the complex null vectors. All the inner products between the tetrad vectors vanish except $l^a n_a = 1 = -m^a \bar{m}_a$. In NP formalism, the metric tensor g_{ij} can be expressed as (cf., Chap. 2)

$$g_{ij} = 2l_{(i}n_{j)} - 2m_{(i}\bar{m}_{j)} \tag{8.42}$$

Using the definition of the spin-coefficients, the covariant derivatives of null vectors are given by ([16] and Chap. 2)

$$\begin{aligned}
l_{i;j} = {}&(\gamma + \bar{\gamma})l_j l_i - \bar{\tau}l_j m_i - \tau l_j \bar{m}_i + (\epsilon + \bar{\epsilon})n_j l_i \\
&-2\bar{\kappa}n_j m_i - \kappa n_j \bar{m}_i - (\alpha + \bar{\beta})m_j l_i - (\bar{\alpha} + \beta)\bar{m}_j l_i \\
&+\bar{\sigma}m_j m_i + \sigma \bar{m}_j \bar{m}_i + \rho m_j \bar{m}_i + \bar{\rho}\bar{m}_j m_i
\end{aligned} \tag{8.43a}$$

$$\begin{aligned}
n_{i;j} = {}&-(\gamma + \bar{\gamma})l_j n_i - \nu l_j m_i - \bar{\nu}l_j \bar{m}_i - (\epsilon + \bar{\epsilon})n_j n_i \\
&+\pi n_j m_i + \bar{\pi}n_j \bar{m}_i + (\alpha + \bar{\beta})m_j n_i + (\bar{\alpha} + \beta)\bar{m}_j n_i \\
&-\lambda m_j m_i - \bar{\lambda}\bar{m}_j \bar{m}_i - \mu \bar{m}_j m_i - \bar{\mu}m_j \bar{m}_i
\end{aligned} \tag{8.43b}$$

$$\begin{aligned}
m_{i;j} = {}&\bar{\nu}l_j l_i - \tau l_j n_i + (\gamma - \bar{\gamma})l_j m_i + \bar{\pi}n_j l_i - \kappa n_j n_i \\
&+(\epsilon - \bar{\epsilon})n_j m_i - \bar{\mu}m_j l_i + \rho m_j m_i + (\bar{\beta} - \alpha)m_j \bar{m}_i \\
&+(\bar{\alpha} - \beta)\bar{m}_j m_i - \bar{\lambda}\bar{m}_j l_i + \sigma \bar{m}_j n_i
\end{aligned} \tag{8.43c}$$

Moreover, from the definition of the differential operators

$$D \equiv l^i \nabla_i, \quad \Delta \equiv n^i \nabla_i, \quad \delta \equiv m^i \nabla_i, \quad \bar{\delta} \equiv \bar{m}^i \nabla_i \tag{8.44}$$

it is easy to write

$$\nabla_i = l_i \Delta + n_i D - \bar{m}_i \delta - m_i \bar{\delta} \tag{8.45}$$

From the definition of the covariant differentiation operators D, Δ, δ and $\bar{\delta}$ along the direction of the vectors of a complex null tetrad, it is possible to write equations (8.43a)–(8.43c) in the following covariant forms:

$$Dl^a = (\epsilon + \bar{\epsilon})l^a - \bar{\kappa}m^a - \kappa\bar{m}^a \tag{8.46a}$$

$$\Delta l^a = (\gamma + \bar{\gamma})l^a - \bar{\tau}m^a - \tau\bar{m}^a \tag{8.46b}$$

$$\delta l^a = (\bar{\alpha} + \beta)l^a - \bar{\rho}m^a - \sigma\bar{m}^a \tag{8.46c}$$

$$\bar{\delta}l^a = (\alpha + \bar{\beta})l^a - \bar{\sigma}m^a - \rho\bar{m}^a \tag{8.46d}$$

$$Dn^a = -(\epsilon + \bar{\epsilon})n^a + \pi m^a + \bar{\pi}\bar{m}^a \tag{8.47a}$$

$$\Delta n^a = -(\gamma + \bar{\gamma})n^a + \nu m^a + \bar{\nu}\bar{m}^a \tag{8.47b}$$

$$\delta n^a = -(\bar{\alpha} + \beta)n^a + \mu m^a + \bar{\lambda}\bar{m}^a \tag{8.47c}$$

$$\bar{\delta}n^a = -(\alpha + \bar{\beta})n^a + \lambda m^a + \bar{\mu}\bar{m}^a \tag{8.47d}$$

$$Dm^a = \bar{\pi}l^a - \kappa n^{a} + (\epsilon - \bar{\epsilon})m^a \tag{8.48a}$$

$$\Delta m^a = \bar{\nu}l^a - \tau n^a + (\gamma - \bar{\gamma})m^a \tag{8.48b}$$

$$\delta m^a = \bar{\lambda}l^a - \sigma n^a + (\beta - \bar{\alpha})m^a \tag{8.48c}$$

$$\bar{\delta}m^a = \bar{\mu}l^a - \rho n^a + (\alpha - \bar{\beta})m^a \tag{8.48d}$$

In this section, we shall translate different types of collineations in the language of NP formalism and obtain a number of results about such collineations. We have the following.

(i) Motion

A symmetry of the spacetime is called motion (M) if the Lie derivative of metric tensor g_{ij} along the vector ξ^a vanishes [cf., Eq. (8.41)], i.e.

$$\pounds_\xi g_{ij} = \xi_{i;j} + \xi_{j;i} = 0 \tag{8.49}$$

Since for null fields, it is possible to choose scaling $l_i \to \varphi l_i$ so that $\gamma = 0$ and $\epsilon = 0$. Thus taking $\xi^i = l^i$ in Eq. (8.49) and using Eq. (8.43a), we get the following theorem.

Theorem 8.1 *Type N pure radiation fields admit motion if and only if $\tau + \bar{\alpha} + \beta = 0$ and l^a is expansion-free.*

Remark 8.1 It may be noted that

  (ii) *Conformal motion* $£_\xi g_{ij} = 2\sigma g_{ij}$,

  (iii) *Special conformal motion* $£_\xi g_{ij} = 2\sigma g_{ij}$, $\sigma_{;jk} = 0$,

  (iv) *Homothetic motion* $£_\xi g_{ij} = 2\sigma g_{ij}$ *i.e. two distances have same constant*
ratios,

all degenerate to motion.

The Ricci tensor for the source-free null electromagnetic fields with propagation
vector l^a is given by [3, 16]

$$R_{ij} = \frac{\chi}{2}\Phi\bar{\Phi}l_i l_j \tag{8.50}$$

It is known that [29, 34] the Weyl tensor can be decomposed into null tetrad
components as

$$\bar{C}_{abcd} = \tfrac{1}{2}(C_{abcd} + iC^*_{abcd})$$

$$= \Psi_0 U_{ab}U_{cd} + \Psi_1(U_{ab}W_{cd} + W_{ab}U_{cd})$$

$$+ \Psi_2(U_{ab}V_{cd} + V_{ab}U_{cd} + W_{ab}W_{cd}) \tag{8.51}$$

$$+ \Psi_3(V_{ab}W_{cd} + W_{ab}V_{cd}) + \Psi_4 V_{ab}V_{cd}$$

where

$$U_{ab} = n_a m_b - m_a n_b,$$

$$V_{ab} = \bar{m}_a l_b - \bar{m}_b l_a, \tag{8.52}$$

$$W_{ab} = l_a n_b - l_b n_a + m_a \bar{m}_b - \bar{m}_a m_b$$

and

$$U_{ab}V^{ab} = 2, \quad W_{ab}W^{ab} = -4 \tag{8.53}$$

so that the Weyl tensor characterizing transverse gravitational field of type N, using
Eqs. (8.10) and (8.52), is given by

$$C_{abcd} = 2Re\Psi V_{ab}V_{cd} \tag{8.54}$$

From Eqs. (8.11a), (8.50) and (8.54), we have

$$F_{ij}l^i = 0, \quad F_{ij}n^i = \tfrac{\Phi}{2}m_j + \tfrac{\bar{\Phi}}{2}\bar{m}_j, \quad F_{ij}m^i = -\tfrac{\bar{\Phi}}{2}l_j, \quad F_{ij}\bar{m}^i = -\tfrac{\Phi}{2}l_j$$

$$R_{ij}l^i = 0, \quad R_{ij}n^i = \tfrac{\chi}{2}\Phi\bar{\Phi}l_j, \quad R_{ij}m^i = 0, \quad R_{ij}\bar{m}^i = 0$$

$$C_{ijkl}l^i = 0, \quad C_{ijkl}n^i = -2Re\Psi\bar{m}_j V_{kl}, \tag{8.55}$$

$$C_{ijkl}m^i = -2Re\Psi l_j V_{kl}, \quad C_{ijkl}\bar{m}^i = 0$$

These equations show that l^i is the common propagation vector for electromagnetic field, Ricci tensor and Weyl tensor.

(v) Ricci collineation

A symmetry of the spacetime is called Ricci collineation (RC) if the Lie derivative of Ricci tensor R_{ij} along vector ξ vanishes, i.e. $\pounds_\xi R_{ij} = 0$.

The Lie derivative of Ricci tensor along the propagation l^a, from the definition, is given by (cf., [3])

$$\pounds_l R_{ij} = D R_{ij} + R_{ik} l^k{}_{;j} + R_{kj} l^k{}_{;i} \tag{8.56}$$

where $D \equiv l^i \nabla_i$. From Eqs. (8.43a), (8.13), (8.46a) and the properties of tetrad $\{l^a, n^a, m^a, \bar{m}^a\}$, Eq. (8.56) reduces to

$$\pounds_l R_{ij} = (\rho + \bar{\rho}) R_{ij} \tag{8.57}$$

Thus, we have the following theorem.

Theorem 8.2 *Type N pure radiation fields admit Ricci collineation if and only if the propagation vector is expansion-free.*

(vi) Free and matter curvature collineations

The Riemann curvature tensor R_{abcd} is defined as

$$R_{abcd} = C_{abcd} + \tfrac{1}{12}(g_{ac} R_{bd} - g_{bc} R_{ad} + g_{bd} R_{ac} - g_{ad} R_{bc})$$
$$+ \tfrac{R}{6}(g_{ad} g_{bc} - g_{ac} g_{bd}) \tag{8.58}$$

while from Eqs. (8.42) and (8.50), for pure radiation fields, we have

$$R = g^{ij} R_{ij} = 0 \tag{8.59}$$

Therefore, for type N PR fields, the curvature tensor is given by

$$R_{abcd} = C_{abcd} + \tfrac{1}{12}(g_{ac} R_{bd} - g_{bc} R_{ad} + g_{bd} R_{ac} - g_{ad} R_{bc})$$
$$= R_{abcd}(F) + R_{abcd}(M) \tag{8.60}$$

where

$$R_{abcd}(F) = C_{abcd} \tag{8.61}$$

denotes the free gravitational part and

$$R_{abcd}(M) = \frac{1}{12}(g_{ac} R_{bd} - g_{bc} R_{ad} + g_{bd} R_{ac} - g_{ad} R_{bc}) \tag{8.62}$$

denotes the matter part of the gravitational field.

Thus, for type N pure radiation fields, from Eqs. (8.42), (8.50) and (8.54), we have

$$R_{abcd}(F) = C_{abcd} = 2Re\Psi V_{ab}V_{cd} \qquad (8.63)$$

and

$$R_{abcd}(M) = \frac{\chi}{4}\Phi\bar{\Phi}(V_{ab}\bar{V}_{cd} + \bar{V}_{ab}V_{cd}) \qquad (8.64)$$

Moreover, from Eqs. (8.52), (8.53) and the properties of tetrad, Eqs. (8.63) and (8.64) lead to

$$R_{abcd}(F)l^a = 0 \quad \text{and} \quad R_{abcd}(M)l^a = 0 \qquad (8.65)$$

which shows that l^a is the propagation vector for the free and matter part of the gravitational field.

The Lie derivative of C_{abcd}, by the definition, is given by

$$\pounds_l C^e_{bcd} = l^h C^e{}_{bcd;h} - C^h{}_{bcd}l^e{}_{;h} + C^e{}_{hcd}l^h{}_{;b} + C^e{}_{bhd}l^h{}_{;c} + C^e{}_{bch}l^h{}_{;d} \qquad (8.66)$$

Now from NP Bianchi identities, the definition of $l^i{}_{;j}$ and the properties of PR fields, we have

$$l^h C^e{}_{bcd;h} = DC^e{}_{bcd} = D\{2Re\Psi V^e{}_b V_{cd}\}$$
$$= 2Re\Psi \rho V^e{}_b V_{cd} \qquad (8.67)$$

$$C^h{}_{bcd}l^e{}_{;h} = 2Re\Psi[(\alpha + \bar{\beta})l_b l^e - \rho l_b \bar{m}^e]V_{cd} \qquad (8.68)$$

$$C^e{}_{hcd}l^h{}_{;b} = 2Re\Psi[-\bar{\tau}l_b l^e - \bar{\rho}\bar{m}_b l^e]V_{cd} \qquad (8.69)$$

etc. With these equations, Eq. (8.66) now reduces to

$$\pounds_l C^e{}_{bcd} = 2Re\Psi[(\rho - \bar{\rho})V^e{}_b - (\alpha + \bar{\beta} + \bar{\tau})l_b l^e + \bar{\rho}m_b l^e]V_{cd}$$
$$= \pounds_l R^e{}_{bcd}(F) \qquad (8.70)$$

Thus, we have the following theorems.

Theorem 8.3 *Type N pure radiation fields admit free curvature collineation (i.e. $\pounds_l R^e_{bcd}(F) = 0$) along the propagation vector l^a if and only if $\bar{\tau} + \alpha + \bar{\beta} = 0$ and $\rho = 0$.*

Since $R^e{}_{bcd} = g^{ea}R_{abcd}$, taking the Lie derivative of Eq. (8.64) along l^a and using Theorems 8.1 and 8.2, we have

Theorem 8.4 *Type N pure radiation fields admit matter curvature collineation (i.e. $\pounds_l R^e_{bcd}(M) = 0$) along the propagation vector l^a if and only if $\tau + \bar{\alpha} + \beta = 0$ and l^a is non-diverging.*

(vii) Curvature collineation

Since $R^e{}_{bcd} = g^{ea} R_{abcd}$, from Eqs. (8.63) and (8.64), (8.60) can be expressed as

$$R^e{}_{bcd} = 2Re\Psi(V^e{}_b V_{cd}) + \frac{\chi}{4}\Phi\bar{\Phi}(V^e{}_b \bar{V}_{cd} + \bar{V}^e_b V_{cd}) \tag{8.71}$$

which on using the properties of tetrad vectors and Eq. (8.52), leads to

$$R^e{}_{bcd} l^c = 0 \tag{8.72}$$

which shows that l^a is the propagation vector for the Riemann curvature tensor. Thus, from Eq. (8.60), Theorems 8.3 and 8.4, we have the following theorem.

Theorem 8.5 *Type N pure radiation fields admit curvature collineation along l^a (i.e. $\pounds_l R^e_{bcd} = 0$) if and only if $\tau + \bar{\alpha} + \beta = 0$ and $\rho = 0$.*

(viii) Affine collineation

The affine collineation (AC) is known to be defined as

$$\pounds_\xi \Gamma^a_{bc} = \xi^a_{;cb} + R^a_{cmb}\xi^m = 0 \tag{8.73}$$

Choosing $\xi^a = l^a$, Eq. (8.73) for pure radiation type N fields, from Eq. (8.72) reduces to

$$\pounds_l \Gamma^a_{bc} = l^a_{;cb} = 0 \tag{8.74}$$

For type N pure radiation fields, using Eqs. (8.43a) and (8.45), the second order covariant derivative of l^a is found to be as

$$\begin{aligned}
l^a_{;cb} = {}&l^a[\{\tau(\alpha + \bar{\beta}) + \bar{\tau}(\bar{\alpha} + \beta)\}l_b n_c\\
&-\{\Delta(\alpha + \bar{\beta}) - 2\gamma(\alpha + \bar{\beta})\}l_b m_c\\
&-\{\bar{\tau}\bar{\pi} + \pi(\alpha + \bar{\beta})\}n_b l_c + \{\rho\pi - D(\alpha + \bar{\beta})n_b m_c\}\\
&-\{\bar{\tau}\bar{\lambda} + \bar{\lambda}(\alpha + \bar{\beta})\}\bar{m}_b l_c\\
&-\{\delta(\alpha + \bar{\beta}) + (\alpha + \bar{\beta})(\beta - \bar{\alpha}) + (\alpha + \bar{\beta})(\bar{\alpha} + \beta) + \rho\lambda\}\bar{m}_b m_c\\
&-\{\bar{\mu}\bar{\tau} - \bar{\mu}(\alpha + \bar{\beta})\}m_b l_c + \rho(\alpha + \bar{\beta})m_b n_c\\
&-\{\bar{\delta}(\alpha + \bar{\beta}) + (\alpha + \bar{\beta})(\alpha - \bar{\beta}) + (\alpha + \bar{\beta})^2 + \bar{\mu}\rho\}m_b m_c]\\
&+n^a[2\tau\bar{\tau}l_b l_c - \rho\bar{\tau}l_b \bar{m}_c + \rho\bar{\tau}m_b l_c - \rho^2 m_b m_c]\\
&+m^a[-(\Delta\bar{\tau} + 2\gamma\bar{\tau})l_b l_c + \{\Delta\rho + \tau^2 + (\alpha + \bar{\beta})\bar{\tau}\}l_b m_c\\
&+(\bar{\alpha} + \beta)\bar{\tau}l_b \bar{m}_c - D\bar{\tau}m_b l_c - \{\delta\bar{\tau} + \bar{\tau}(\bar{\alpha} + \beta)\}\bar{m}_b l_c\\
&+\{\bar{\tau}\bar{\rho} + \bar{\rho}(\alpha + \bar{\beta})\}\bar{m}_b m_c\\
&-\{\bar{\delta}\bar{\tau} + \bar{\tau}(\alpha + \bar{\beta}) + \bar{\tau}(\alpha - \bar{\beta})\}m_b l_c + \rho(\alpha - \bar{\beta})m_b m_c + \bar{\tau}\rho m_b \bar{m}_c]\\
&+\bar{m}^a[-\rho\tau l_b n_c + \tau\bar{\tau}l_b m_c + \rho\pi n_b l_c + D\rho n_b m_c\\
&+\{\delta\rho + \rho(\beta - \bar{\alpha}) + \rho(\bar{\beta} - \alpha)\}\bar{m}_b m_c + \rho\bar{\lambda}\bar{m}_b l_c\\
&+\rho\bar{\mu}m_b l_c - \rho^2 m_b n_c + \{\rho(\alpha + \bar{\beta}) + \rho\delta + \rho(\alpha - \bar{\beta})\}m_b m_c]\\
&+ c.\,c.
\end{aligned} \tag{8.75}$$

Thus, from Eqs. (8.74) and (8.75), we have the following theorem.

Theorem 8.6 *The necessary and sufficient conditions for type N pure radiation fields to admit affine collineation are that*

$$\alpha + \bar{\beta} = 0, \rho = 0, \tau = 0 \tag{8.76}$$

(ix) Projective collineation

It is known that a projective collineation (PC) along l^a is defined as

$$\pounds_l \Gamma^a{}_{bc} = \delta^a{}_b A_{;c} + \delta^a{}_c A_{;b} \tag{8.77}$$

where A is an arbitrary function. Now

$$A_{;c} = A_{;i} g_c^i = (l_i \Delta + n_i D - \bar{m}_i \delta - m_i \bar{\delta}) A g_c^i$$

$$= (l_i \Delta + n_i D - \bar{m}_i \delta - m_i \bar{\delta}) A (l^i n_c + n^i l_c - m^i \bar{m}_c - \bar{m}^i m_c) \tag{8.78}$$

$$= \Delta A l_c + D A n_c - \delta A \bar{m}_c - \bar{\delta} A m_c$$

which shows that $A_{;c} = 0$ if and only if

$$DA = \Delta A = \delta A = \bar{\delta} A = 0 \tag{8.79}$$

That is, A is constant.

Moreover, from Eqs. (8.73) and (8.77), we have

$$\xi^a_{;cb} + R^a_{cmb} \xi^m = \delta^a{}_b A_{;c} + \delta^a{}_c A_{;b} = 0 \tag{8.80}$$

Choosing $\xi^a = l^a$, since for type N pure radiation fields we have $R^a_{cmb} l^m = 0$ [cf., Eq. (8.72)], therefore Eq. (8.80) yields

$$l^a_{;bc} = \delta^a{}_b A_{;c} + \delta^a{}_c A_{;b} = 0 \tag{8.81}$$

These tensorial equations are equivalent to

$$\alpha + \bar{\beta} = 0, \quad \tau = 0, \quad \rho = 0$$
$$DA = \Delta A = \delta A = \bar{\delta} A = 0 \tag{8.82}$$

Thus, we have the following theorems.

Theorem 8.7 *For type N pure radiation fields, projective collineation degenerate to affine collineation.*

Remark 8.2 In a similar way we can show that for type N pure radiation fields, the following collineations degenerate to affine collineation.

(x) Special projective collineation (SPC)

$$\pounds_l \Gamma^a_{bc} = \delta^a_b A_{;c} + \delta^a_c A_{;b}, \quad A_{;bc} = 0 \tag{8.83}$$

where A is arbitrary function.
(xi) Conformal collineation (Conf C)

$$\pounds_l \Gamma^a_{bc} = \delta^a_b B_{;c} + \delta^a_c B_{;b} - g_{bc} g^{ad} B_{;d} \tag{8.84}$$

where B is arbitrary function.
(xii) Special conformal collineation (S Conf C)

$$\pounds_l \Gamma^a_{bc} = \delta^a_b B_{;c} + \delta^a_c B_{;b} - g_{bc} g^{ad} B_{;d}, \quad B_{;bc} = 0 \tag{8.85}$$

(xiii) Null geodesic collineation (NC)

$$\pounds_l \Gamma^a_{bc} = g_{bc} g^{ad} E_{;d} \tag{8.86}$$

where E is arbitrary function.
(xiv) Special null geodesic collineation (SNC)

$$\pounds_l \Gamma^a_{bc} = g_{bc} g^{ad} E_{;d}, \quad E_{;bc} = 0 \tag{8.87}$$

The double covariant derivative of A, using Eqs. (8.45)–(8.48), is given by

$$A_{;b;c} \equiv A_{;bc} = [-(\mu + \bar{\mu})n_b l_c + (\bar{\rho} - \rho)n_b n_c + \{-\bar{\tau} + \pi + (\alpha - \bar{\beta})\}n_b m_c]DA$$

$$+ [-(\mu + \bar{\mu})l_b l_c + (\bar{\rho} + \rho)l_b n_c + \{-\bar{\tau} + \pi - (\alpha - \bar{\beta})\}l_b m_c]\Delta A$$

$$+ [(\mu + \bar{\mu})\bar{m}_b l_c + (\bar{\rho} - \rho)\bar{m}_b n_c + \{\bar{\tau} - \pi + (\alpha - \bar{\beta})\}\bar{m}_b m_c$$

$$+ \{\tau - \bar{\pi} + (\alpha - \bar{\beta})\}\bar{m}_b \bar{m}_c]\delta A$$

$$+ c.\, c. \tag{8.88}$$

Finally, we have the following.

(xv) Maxwell Collineation (MC)

A spacetime is said to admit Maxwell collineation [13] if there exists a vector field ξ^a such that

$$\pounds_\xi F_{ij} = F_{ij;k}\xi^k + F_{ik}\xi^k_{;j} + F_{kj}\xi^k_{;i} = 0 \tag{8.89}$$

The geometrical symmetry defined by Eq. (8.89) along with the symmetry given by Eq. (8.49) has been the subject of interest for quite some time. Thus, for example, for non-null electromagnetic fields, Woolley [33] has shown that if Eq. (8.49) holds then F_{ab} satisfies $\pounds_\xi F_{ab} = k(\alpha) F_{ab}$ form some constant $k(\alpha)$, $\alpha = 1, 2, \ldots r$; while Michalski and Wainwright [24] have shown that $\pounds_\xi g_{ab} = 0$ implies $\pounds_\xi F_{ab} = 0$ for non-null fields. On the other hand, for non-null fields, Duggal [17] has proved the converse part under certain conditions. Maxwell collineations have also been studied by Ahsan [1, 3] and Ahsan and Hussain [10]. It is seen that for null electromagnetic fields neither MC is a consequence of Motion nor Motion is a consequence of Maxwell collineation. Moreover, using Newman–Penrose formalism, Ahsan [3] has obtained the conditions under which a null electromagnetic field may admit Maxwell collineation and Motion. The concept of Maxwell collineation was further extended as Maxwell Inheritance (MI) by Ahsan and Ahsan [7], who applied this concept to (i) the spacetime solution corresponding to strong gravitational waves propagating in generalized electromagnetic universes [23] and (ii) the algebraically general twist-free solution of Einstein–Maxwell equation for non-radiative electromagnetic fields [32].

Choosing $\xi^a = l^a$ (l^a is considered as the propagation vector) and using Eqs. (8.11a), (8.43a) and (8.46)–(8.48), Eq. (8.89) leads to the following.

Theorem 8.8 *Type N pure radiation fields satisfying source-free Maxwell equations always admit Maxwell collineation along the propagation vector.*

Remark 8.3 From Theorems 8.1, 8.2 and 8.8, it may be noted that the spacetime solution (8.40) representing the interaction of pure radiation fields always admits Maxwell collineation but admits Motion and Ricci collineation under certain conditions.

References

1. Ahsan, Z.: Tamkang J. Math. **9**(2), 237–240 (1978)
2. Ahsan, Z.: J. Math. Phys. Sci. **21**(5), 515–526 (1987)
3. Ahsan, Z.: Acta Phys. Sin. **4**(5), 337–343 (1995)
4. Ahsan, Z.: Braz. J. Phys. **26**(3), 572–576 (1996)
5. Ahsan, Z.: Indian J. Pure Appl. Math. **31**(2), 215–225 (2000)
6. Ahsan, Z.: Bull. Cal. Math. Soc. **97**(3), 191–200 (2005)
7. Ahsan, Z., Ahsan, N.: Bull. Cal. Math. Soc. **94**(5), 385–388 (2002)
8. Ahsan, Z., Ali, M.: Int. Jour. Theo. Phys. **51**, 2044–2055 (2012)
9. Ahsan, Z., Ali, M.: Int. J. Theo. Phys. **54**(5), 1397–1407 (2015)
10. Ahsan, Z., Hussain, S.I.: Ann. di Mat. Pur. ed Applicata CXXXVI, 379–396 (1980)
11. Ali, M., Ahsan, Z.: Global J. Adv. Res. Class. Modern Geom. **1**(2), 76–85 (2012)
12. Ali, M., Ahsan, Z.: Proceedings, International Conference D.G. Functional Analysis and Applications, Jamia Millia Islamia, New Delhi, editor - Hasan Shahid, M. et al. (Ed.), pp. 1–10 (2014) ISBN: 978-81-8487-421-1, 1-10
13. Collinson, C.D.: Gen. Relat. Gravitat. **1**, 137–139 (1970)
14. Davis, W.R., Green, L.H., Norris, L.K.: Il Nuovo Cimento **34B**, 256–280 (1976)
15. Davis, W.R., Moss, M.K.: Il Nuovo Cimento **65B**, 19–32 (1970)
16. Debney, G.C., Zund, J.D.: Tensor (N.S.), **22**, 333–340 (1971)

17. Duggal, K.L.: Existence of two Killing vector fields on the spacetime of general relativity. Preprint (1978)
18. Duggal, K.L.: J. Math. Phys. **28**, 2700–2705 (1987)
19. Duggal, K.L., Sharma, R.: J. Math. Phys. **27**, 2511–2514 (1986)
20. Katzin, G.H., Livine, J., Davis, W.R.: J. Math. Phys. **10**(4), 617–630 (1969)
21. Katzin, G.H., Livine, J., Davis, W.R.: J. Math. Phys. **11**, 1875 (1970)
22. Katzin, G.H., Levine, J.: J. Colloq. Math. **26**, 21–38 (1972)
23. Khlebnikov, V.I.: Class. Quant. Grav. **3**(2), 169–173 (1986)
24. Michalaski, H., Wainwright, J.: Gen. Relativ. Gravit. **6**(3), 289 (1975)
25. Newman, E.T., Penrose, R.: J. Math. Phys. **3**, 566–579 (1962)
26. Norris, L.K., Green, L.H., Davis, W.R.: J. Math. Phys. **18**, 1305–1312 (1977)
27. Oliver, D.R., Davis, W.R.: Gen. Relativ. Gravit. **8**, 905–914 (1977)
28. Parga, G.A., Lobez Bonilla, J.L., Ovando, G., Chassin, M.: Rev. Mex. Fisica **35**, 393 (1989)
29. Stephani, H., Krammer, D., McCallum, M., Herlt, E.: Exact Solutions of Einstein's Field Equations, 2nd edn. Cambridge University Press, Cambridge (2003)
30. Szekeres, P.: J. Math. Phys. **6**, 1387 (1965)
31. Talbot, J.: Comm. Math. Phys. **13**, 45 (1969)
32. Tariq, N., Tupper, B.O.J.: Class. Quant. Grav. **6**, 345 (1975)
33. Woolley, M.L.: Comm. Math. Phys. **31**(1), 75–81 (1973)
34. Zakharov, V.D.: Gravitational Waves in Einstein Theory, Halsted Press. John Wiley & sons, New York (1973)

Concluding Remarks

The correspondence between electromagnetism and gravitation is very rich and detailed. Some of these correspondences are still uncovered, while some of them are further developed. This correspondence is reflected in Maxwell-like form of the gravitational field tensor (the Weyl tensor), the super energy–momentum tensor (the Bel–Robinson tensor) and the dynamical equations (the Bianchi identities). It is also known that an electromagnetic field can always be generated through a vector potential A_i. In an analogy to electromagnetism, can we generate a gravitational field through a potential (that should be tensorial in nature)? The answer is in affirmative—indeed it is possible to generate the gravitational field (the Weyl tensor) through the process of covariant differentiation of a third rank tensor L_{ijk}. This is what precisely done by Cornelius Lanczos (a Hungarian mathematician and physicist) in 1962. This tensor is now commonly known as Lanczos potential or Lanczos spin tensor. Unfortunately, this potential exists only in four dimensions and there is no such potential for the Riemann tensor when the space is not Ricci-flat, i.e. $R_{ij} \neq 0$. However, there are several reasons that why the study of such potential is important and some of them are

  (i) definition of energy and momentum,
 (ii) possibility of 'massive gravitons', then the potential becomes dynamic,
(iii) quantization,
(iv) dealing with a simpler object.

The equations that provide the relationship between this tensor and the Weyl tensor are known as Weyl–Lanczos equations. For a given spacetime geometry, the construction of Lanczos potential is equivalent to solving Weyl–Lanczos equations under certain constraint conditions imposed on L_{ijk}. There are several ways of solving Weyl–Lanczos equations, although none of them are as straightforward as one would like them to be. However, the tetrad formalisms offer some simplifications and this is what we have done in this text.

Using GHP formalism, the solutions of Weyl–Lanczos equations, which in turn leads to Lanczos potential, have been obtained for arbitrary Petrov types II and D

© Springer Nature Singapore Pte Ltd. 2019
Z. Ahsan, *The Potential of Fields in Einstein's Theory of Gravitation*,
https://doi.org/10.1007/978-981-13-8976-4

spacetimes. The results obtained are supported by examples, and it is seen that the Lanczos potential for Robinson–Trautman metric of Petrov type II depends upon the radial coordinate r; while for Kerr black hole, the Lanczos potential is related to the mass parameter of the Kerr black hole and the Coulomb component of the gravitational field. While using NP formalism, a general prescription to obtain the Lanczos potentials for radiative spacetimes (Petrov types III and N) have also been given and consequently the Lanczos potential for the well-known radiative solutions of Einstein field equations have been calculated. In this way, we have found a general prescription for obtaining the Lanczos potential for algebraically special spacetimes (Petrov types II, D, III and N).

Using the method of general observers and the spin-coefficient formalism of Newman and Penrose, a yet another method for obtaining the Lanczos potential for perfect fluid spacetimes has been given. The kinematical quantities and the equations satisfied by them have been translated into NP formalism, and in the process, the Lanczos potential for shear-free irrotational perfect fluid spacetimes and Bianchi Type I spacetimes have been obtained. As an example to Bianchi Type I models, the Lanczos potential for Kasner metric has been found and it is seen that the Lanczos scalars depend upon time and the constants appearing in the Kasner metric.

The Lanczos potential for some well-known solutions of Einstein and Einstein–Maxwell equations have also been obtained using the techniques of tetrad formalisms. It has been observed that the Lanczos scalars can be expressed in terms of the spin-coefficients, and our conjecture is that it shall occur in any Petrov type if we select an adequate null tetrad. Moreover, since Lanczos spin tensor is a geometrical object of spacetime, therefore it can be interpreted physically, and an attempt has been made to assign a possible physical meaning to this tensor. Thus, for example, in case of Gödel spacetime, the Lanczos potential depend upon the parameter that is responsible for the rotation of the fluid. While for a rotating black hole, the Lanczos potential depends upon the mass of the black hole and Coulomb component of the gravitational field. For Schwarzschild exterior solution and Vaidya's solution for the external field of a radiating star, the Lanczos scalars are inversely proportional to the radial distance. Also since Petrov type D fields have only Coulomb component Ψ_2 of the gravitational field with l^i and n^i as the propagation vectors therefore Lanczos scalars can act as the potential of the gravitational field, and thus justifying the name—the Lanczos potential. The non-uniqueness character of Lanczos potential has also been established and it is seen that this character has been achieved by the different choices of the tetrad vectors. This non-uniqueness property of the Lanczos potential is in close analogy with the potential of the electromagnetic field.

Apart from tetrad formalism, there are other methods for the study of Lanczos potential and we have discussed these methods to obtain the Lanczos potential of the Gödel cosmological model. It is seen that the Gauss equation employed in the embedding of a four-dimensional Riemannian manifold into a five-dimensional Euclidean space allows the existence of a symmetric tensor which in turn generates the Lanczos potential. Thus, a connection between the embedding of four-dimensional Riemannian manifold and the Lanczos potential has been established. Using the method of the Lovelock's theorem, the Lanczos potential for the Gödel cosmological model is obtained and a possible physical meaning is assigned to the Lanczos potential for this model. In fact, the Lanczos potential of the Gödel cosmological model represents some type of angular momentum. Considering a second rank symmetric tensor satisfying the wave equation for the Gödel spacetime, a solution of the wave equation has been obtained, which in turn generates the Lanczos potential for this cosmological model.

There are still some areas where the Lanczos potential can be studied—for example, if there is a possibility of 'massive gravitons', then what will happen to Lanczos potential? Will it become dynamic (as in the case of electromagnetic field where the vector potential is dynamic)? Whether or not the Lanczos potential can be used for the quantization of the gravitational field. Moreover, there is no general method available to find the Lanczos potential of an arbitrary Petrov type I (algebraically general) gravitational fields.

The interaction of a Petrov type N gravitational field with a null electromagnetic field has been considered and a metric describing such a situation has been obtained using Newman–Penrose formalism. The geometric and physical properties of the solution have also been discussed. As another application of NP formalism, the geometrical symmetries corresponding to the continuous groups of motions generated by a null vector have been considered and for Petrov type N pure radiation fields, these symmetries have been studied in detail.

Printed in the United States
By Bookmasters